PA D

P9-BYI-902

JEFFREY J. CONTELMO, P.E.
P.O. BOX 526
HUGHSONVILLE, NY 12537

APPLIED MECHANICS FOR ENGINEERING TECHNOLOGY

SECOND EDITION

APPLIED MECHANICS FOR ENGINEERING TECHNOLOGY

SECOND EDITION

Keith M. Walker
Professor of Engineering
Red River Community College
Winnipeg, Manitoba, Canada

RESTON PUBLISHING COMPANY, INC.
A Prentice-Hall Company
Reston, Virginia

Library of Congress Cataloging in Publication Data

Walker, Keith M
   Applied mechanics for engineering technology.

   Includes index.
   1.   Mechanics, Applied.   I.   Title.
TA350.W19 1978   620.1′05   77-28797
ISBN 0-87909-025-1

©1978 by
RESTON PUBLISHING COMPANY, INC.
Reston, Virginia 22090
A Prentice-Hall Company

10 9 8 7 6 5 4 3 2

Printed in the United States of America.

TO MY WIFE JUDY

who gave support in many ways

CONTENTS

II. DYNAMICS

PREFACE

Applied Mechanics is more than the teaching of physics principles. It is an important instrument in developing a method of stripping a problem to essentials and solving in a logical, organized manner. This method of working can be applied to many other areas. This book, therefore, shows a consistent pattern of problem solving. The physics principles are presented in small elementary steps, the mathematics is kept at a reasonable level, and the problems are as practical as possible without becoming too involved with many extraneous details.

To accommodate the transition years between the English system and the SI metric system, each chapter is a random mix of both systems but predominantly SI metric.

There are more than 140 worked examples and 650 graded problems of which nearly two-thirds are in the SI metric system.

Chapters 1 through 9 cover Statics and Chapters 10 through 15 cover Dynamics. The appendix includes graphical solutions, Bow's notation, Maxwell's diagrams, and steel tables.

The answers which are provided for the odd-numbered problems coupled with the informal approach used in this book should promote self-study and problem solving by the student.

Keith M. Walker

STATICS

1

INTRODUCTION

1-1 WHAT AND WHY OF APPLIED MECHANICS

To someone who has never been exposed to applied mechanics, the subject may seem on first examination to be closely akin to a formal physics course since physics is what it most closely relates to in the high-school curriculum. But applied mechanics is basically an engineering science with practical applications. This text will not emphasize the purely theoretical approach but will endeavor to show the practical applications of new theory.

Basic mechanics is composed of two principal areas—statics and dynamics. In this book, *statics* will be dealt with first; it is the study of forces on and in structures that are at rest or moving at a uniform velocity. A motionless body may have several forces acting on it, e.g., gravitational force and a force opposing that gravity. Such a body is therefore *static*, or motionless, and has forces in balance, or *equilibrium*. Statics is the analyzing and determining of such forces. *Dynamics*, which will be studied later, is the next logical step in the study of forces since it is concerned with *dynamic equilibrium*, or the forces acting on a moving body.

Applied mechanics, since it deals with the very basic concept of force, is the origin for all calculations in areas such as stress analysis, machine design, hydraulics, and structural design. The design of an aircraft landing gear would require knowledge in all of these areas.

There are reasons other than the above for learning mechanics: the discipline is invaluable in developing one's logic or reasoning ability; one also learns a method of applying a little theory in a logical, neatly organized manner to arrive at a solution to a practical problem.

The key to success is the method of attacking problems rather than the learning of massive quantities of theory. For those who prefer to memorize equations and to look for a "plug into the formula" solution, a change in method will be required.

The "why" of applied mechanics is therefore twofold: to lay the groundwork of theory for future engineering calculations and to train a person to organize and present his work in a logical manner. Such theory and the logical thought processes that must accompany it are the groundwork for many future engineering subjects.

1-2   UNITS AND BASIC TERMS

The units in this book will be predominantly SI metric; the remainder will be the English system.

A metric system was standardized in June, 1966 when the International Organization for Standardization approved a metric system called *Le Systeme International d'Unites*. The abbreviation is SI. This supplanted the old MKS metric system.

There are some changes required in the previous metric system, but the most marked change will be for countries converting from the English system to the SI metric system. Due to the current phase of conversion to the SI system, the units used in examples and problems will be randomly mixed throughout the book.

In the SI system, there are only six basic units (Table 1-1).

Table 1-1

PHYSICAL QUANTITY	NAME	SYMBOL
length	meter	m
mass	kilogram	kg
time	second	s
electric current	ampere	A
temperature	Kelvin	K
luminous intensity	candela	cd

These basic units must measure quantities that could vary considerably in magnitude. To avoid awkwardly large or small figures, prefixes representing multiples and submultiples will be used (Table 1-2).

Table 1-2

NAME	SYMBOL	MULTIPLY BY
MULTIPLES		
kilo	k	10^3
mega	M	10^6
giga	G	10^9
tera	T	10^{12}
SUBMULTIPLES		
milli	m	10^{-3}
micro	μ	10^{-6}
nano	n	10^{-9}
pico	p	10^{-12}

You will notice that the multiples and submultiples are in increments of three digits. There are others that do not follow this pattern and therefore are not part of the SI system. Their use is permitted for convenience in certain cases.

The multiples are:

hecto	h	multiply by 10^2
deka	da	multiply by 10

Table 1-3

QUANTITY	UNIT	SYMBOL	DESCRIPTION
Acceleration	Meter per second squared	—	m/s^2
Angle	Radian	rad	—
Angular acceleration	Radian per second squared	—	rad/s^2
Angular momentum	Kilogram meter squared per second	—	$kg \cdot m^2/s$
Angular velocity	Radian per second	—	rad/s
Area	Square meter	—	m^2
Density	Kilogram per cubic meter	—	kg/m^3
Energy	Joule	J	$N \cdot m$
Force	Newton	N	$kg \cdot m/s^2$
Frequency	Hertz	Hz	s^{-1}
Length	Meter	m	—
Mass	Kilogram	kg	—
Moment (Torque)	Newton-meter	—	$N \cdot m$
Momentum	Kilogram meter per second	—	$kg \cdot m/s$
Power	Watt	W	J/s
Pressure	Pascal	Pa	N/m^2
Strain	—	—	mm/mm
Stress	Pascal	Pa	N/m^2
Time	Second	s	—
Velocity	Meter per second	—	m/s
Volume			
Solids	Cubic meter	—	m^3
Liquids	Liter	ℓ	$10^{-3}m^3$
Work	Joule	J	$N \cdot m$

The submultiples are:

deci	d	multiply by 10^{-1}
centi	c	multiply by 10^{-2}

For those who may not be familiar with the metric prefixes, the following equivalent values will demonstrate their use.

$$1 \text{ kilometer} = 1000 \text{ meters}$$
$$1 \text{ km} = 1000 \text{ m}$$

$$1 \text{ millimeter} = 10^{-3} \text{ meters}$$
$$1 \text{ mm} = 10^{-3} \text{ m}$$
$$10^3 \text{ mm} = 1 \text{ m}$$

Some of the principal SI-derived units are shown in Table 1-3.

The units and terms in Table 1-3 will be discussed when each specific area is covered. The following discussion of each will serve as an introductory explanation and as a central reference.

Length

A base unit of 1 meter (m) is used. The popular multiples are kilometers (km) and millimeters (mm). The centimeter (cm) is used for calculations to avoid unwieldy numbers and for convenience in other cases. The predominant unit in the English system is the foot (ft).

Mass

The *mass* of an object is a measure of the amount of material in the object. A base unit of 1 kilogram (kg) is used (1 tonne = 1000 kg). In the English system, it is the slug. Mass, weight, and force of gravity are discussed further under the heading of "Force."

Time

The base unit of time is 1 second (s). Note that the abbreviation is "s" rather than "sec" as in the English system. Because of universal acceptance, other permitted units are minute (min), hour (h), and day (d).

Area

Area is measured in square meters (m^2) or multiples such as square millimeters (mm^2), square centimeters (cm^2), and square kilometers (km^2). Area in the English system is often in ft^2 or yd^2.

Volume

The base unit for solids is 1 cubic meter (m^3), and for liquids it is 1 liter (ℓ), which is equal to 1 cubic decimeter (dm^3). Another common relationship between solids and liquids is 1 milliliter (ml) = 1 cubic centimeter (cm^3). The English system commonly uses ft^3 and yd^3 for solids and gallons for liquid.

Force

The unit of force is the newton (N). One newton is the force that when applied to a mass of 1 kg gives it an acceleration of 1 m/s^2 (1 $N = 1 kg \cdot m/s^2$). Similarly, a mass of 1 kg with the standard acceleration of gravity of 9.81 m/s^2 will have a force of gravity of $1 \times 9.81 = 9.81$ N.

In the English system, a mass of 1 slug has a weight or force of gravity = mass × acceleration of gravity = 1 slug × 32.2 $ft/sec^2 = 32.2$ lb. Note that in the SI system the term "weight" of an object is not usually used, but rather the force of gravity expressed in newtons. The newton is a relatively small unit of force; therefore, common multiples are kN and MN. To handle large forces in the English system, the kilopound (kip) is used (1 kip = 1000 lb).

Angle

The *radian* (rad) is used for measuring plane angles. In common practice, plane angles will continue to be measured in degrees although the use of minutes and seconds is discontinued, e.g., 38.2° rather than 38° 12′.

The radian is the angle between two radii of a circle that cut off on the circumference an arc equal in length to the radius. Since the circumference = $2\pi r$, there are 2π rad. in 360°.

Pressure

Pressure is force per unit area, and the derived unit used is 1 pascal = 1 newton per square meter (1 $Pa = 1 N/m^2$). Again, this is a relatively small unit; therefore, kPa and MPa are often used. Units of psi ($lb/in.^2$) are common in the English system.

Stress

Stress is an internal load per unit area; therefore, it is expressed in units of pascals as is pressure.

Strain

Strain is a measure of length per length, and the SI system requires the use of millimeters/millimeter (mm/mm). Units of cm/cm may be encountered since they are still in common usage in some countries. The English system uses units of in./in.

Energy

The *joule* is the work done when a force of 1 newton acts through a distance of 1 meter. Since

$$\text{work} = \text{force} \times \text{distance}$$
$$1 \text{ J} = 1 \text{ N} \cdot \text{m}$$

Similar to the pascal, the joule is a relatively small unit and may often be preceded by the larger prefixes of kilo and mega. In the English system, a force of 1 pound acts through a distance of 1 foot giving units of ft-lb.

Work

Since work is energy, it has units of joules.

Power

Power is the rate of doing work. One *watt* of power is the rate of 1 joule of work per second.

$$\text{power} = \frac{\text{work}}{\text{time}}$$
$$1 \text{ W} = 1 \text{ J/s} = 1 \text{ N} \cdot \text{m/s}$$

All forms of power are expressed in watts. An electric motor has an electrical power input of watts and a mechanical power output of watts.

In the English system, power is in horsepower

$$\text{power} = \frac{\text{work}}{\text{time}} = \frac{\text{ft-lb}}{\text{sec}}$$

but   1 horsepower (hp) = 550 ft-lb/sec

Therefore $$\text{hp} = \frac{\text{ft-lb/sec}}{550}$$

Moment

Moment is equal to a force times a perpendicular distance and is expressed in units of N·m or multiples thereof. In the English system, units of lb-ft are used to make a distinction from work units expressed in ft-lb.

Velocity

Since *velocity* is a rate of change of displacement with respect to time, the units are m/s. Usual English system units are ft/sec, ft/min, and miles per hour (mph).

Angular Velocity

Angular velocity is the rate of change of rotational displacement with respect to time and is expressed in units of radians per second (rad/s). Revolutions per minute (rpm) is a permitted term but is converted to rad/s for calculations. This applies to both SI and the English systems.

Acceleration

Acceleration is the rate at which velocity changes; hence, the units are (m/s)/s or m/s². The commonly used unit in the English system is ft/sec². The acceleration of gravity in each system is 9.81 m/s² and 32.2 ft/sec².

Angular Acceleration

Angular acceleration is the rate at which angular velocity changes; hence, the units are (rad/s)/s or rad/s². The English system is similar.

Density

Since *density* is mass per unit volume, the units are kilogram per cubic meter (kg/m^3). The English system usually uses $lb/in.^3$ or lb/ft^3.

Frequency

One hertz is the frequency of a periodic occurrence that has a period of 1 second. Formerly used units were cycles per second ($1\ Hz = 1\ s^{-1}$). This also applies to the English system.

Momentum

Momentum is mass times velocity; it is expressed in units of $kg \times m/s = kg \cdot m/s$. In the English system, we have mass in slugs or $(lb\text{-}sec^2)/ft$ times velocity in ft/sec giving units of lb-sec.

Angular Momentum

Angular momentum is mass moment of inertia times angular velocity; it is expressed in units of $kg \cdot m^2 \times rad/s = kg \cdot m^2/s$. In the English system, we have $ft\text{-}lb\text{-}sec^2 \times rad/sec = ft\text{-}lb\text{-}sec$.

For those who are not familiar with the writing of the various SI symbols, the following rules may be useful.

1. Symbols of units named after historic persons are written in capital letters, e.g., M, G, and T. All others are written in lower-case letters.
2. Symbols are not written with a plural "s."
3. Symbols are never followed by a period.
4. Compound prefixes cannot be used; for example, $m\mu m$ should be written nm.
5. Avoid the use of a prefix in the denominator of a composite unit; for example, do not use N/mm, but rather kN/m.

Table 1-4 provides some conversion factors that you may need while using this book in order to make conversions between the English system and the SI system. For those converting from English to SI, let me again emphasize that the key to learning the SI metric units is to think and work in the quantities of the system and not to have to convert continually from the English system: *Think metric!*

Table 1-4

Length	1 in	= 25.4 mm
	1 ft	= 0.3048 m
	1 mile	= 1609 m
Area	1 in²	= 6.45 cm²
	1 ft²	= 0.093 m²
	1 sq mile	= 2.59 km²
Volume	1 in³	= 16.39 cm³
	1 ft³	= 0.0283 m³
Capacity	1 qt	= 1.136ℓ
	1 gal	= 4.546ℓ
Mass	1 lb	= 0.454 kg
	1 slug	= 14.6 kg
Velocity	1 in/sec	= 0.0254 m/s
	1 ft/sec	= 0.3048 m/s
	1 ft/min	= 0.00508 m/s
	1 mph	= 0.447 m/s = 1.61 km/h
Acceleration	1 in/sec²	= 0.0254 m/s²
	1 ft/sec²	= 0.3048 m/s²
Force	1 lb	= 4.448 N
	1 poundal	= 0.138 N
Pressure	1 lb/in²	= 6.895 kPa
	1 lb/ft²	= 47.88 Pa
Energy	1 ft-lb	= 1.356 J
	1 BTU	= 1.055 kJ
	1 hp-hr	= 2.685 MJ
	1 watt-hr	= 3.6 kJ
Power	1 hp	= 0.746 kW

1-3   METHOD OF PROBLEM SOLUTION AND WORKMANSHIP

In initially trying to understand and analyze a problem, the student should attempt to utilize his intuition, experience, and visualization. The use of common sense can be quite helpful here.

Once the solution is begun—with the use of free-body diagrams (Section 4-2), for example—there are general rules and equations to be used for the remainder of the solution. In this way, nothing is left to chance. (A free-body diagram is a diagram of the object showing various forces acting on it.)

Upon obtaining an answer, the student may know from experience that a check is required. In many cases, an alternate method may be used as a reliable check. The checking of calculations by the same method can often lead to the same mathematical error; therefore, the use of an alternate method, where possible, is advisable.

A typical problem solution should begin with a reproduction of the given or known information in a concise form. While this may seem wasteful of time, it serves several useful purposes:

1. it thoroughly acquaints one with the problem;
2. it is typical preliminary organization practiced every day in engineering; and
3. the problem can be easily reviewed later without referring to the text or other sources.

The suggested method of solution consists of diagrams on the left side of the page and calculations on the right side. Do not crowd the calculations or double back with them. The format of the Example 1-1 demonstrates what has just been described.

It is suggested that any symbols or methods of presentation used in this text be adhered to. Any deviation or short-cut method might be successful at the early stages but could lead to complications in more sophisticated problems later. It is better to do this now than to wish later that you had.

It cannot be overemphasized that, through all phases of problem solution, neatness is an important factor contributing to organization. This will be realized when one refers back for intermediate answers or checks the final answer.

Although at this point it is premature to expect that the student understand all the conventions and equations used, Example 1-1 will be solved on page 14. A file of problems solved in this manner might be all that a student would require for review or studying purposes at the end of the course.

Example 1-1

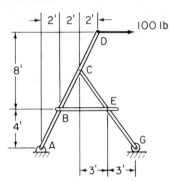

Figure 1-1a

A force of 100 lb is applied to the frame shown. Calculate the load in member BE.

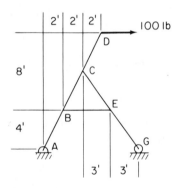

Figure 1-1b

Given: As shown

Required: BE

(See the following page for a description of each step.)

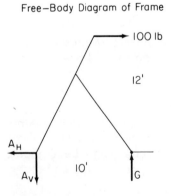

Free—Body Diagram of Frame $\Sigma M_A = 0$

$$G \times 10 = 100 \times 12$$
$$\underline{G = 120 \text{ lb} \uparrow}$$

Figure 1-1c

Free–Body Diagram of CG $\Sigma M_C = 0$

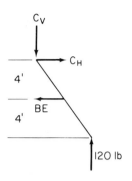

$$BE \times 4 = 120 \times 6$$

$$BE = \frac{120 \times 6}{4}$$

$$BE = 180 \text{ lb} \leftarrow$$

$$\underline{BE = 180 \text{ lb tension}}$$

Figure 1-1d

By referring to the solved Example 1-1, you will note that the solution was arranged on the page in the following manner:

1. Sketch of object, frame, etc., reproduced in a concise form suitable for reference during problem solution.	*2.* State given information not already labelled on sketch.
	3. State required information.
4. Free-body diagram or further sketch.	*5.* Statement of type of calculation to be used.
	6. Calculations.
	7. Underlined answer to calculations.
8. Further free-body diagram if necessary.	*9.* Statement, calculations, and answer pertaining to 8.
	10. Final answer underlined.

1-4 NUMERICAL ACCURACY AND SIGNIFICANT FIGURES

The way a quantity or number is written indicates a *degree of accuracy*. If you measured a distance of 45 ft and recorded it as 45.0 ft, then you would be implying an accuracy of plus or minus 0.05 ft. You must be careful not to indicate any greater accuracy than is intended or possible.

The final answer of a calculation must not indicate any greater accuracy than that of the most inaccurate figure of the original data.

Suppose that we perform the calculation $27.1 \times 89.2/203$. The indicated accuracy for each figure is as follows:

$$27.1 \pm 0.05 = \pm 0.18\%$$
$$89.2 \pm 0.05 = \pm 0.06\%$$
$$203 \pm 0.5 = \pm 0.25\%$$

The most inaccurate figure is 203 plus or minus 0.5, giving an accuracy of approximately ± 0.25 percent. Performing the complete calculation, we get 11.90798. Stating the answer as 11.9 implies an accuracy of $0.05/11.9 \times 100 = 0.42\%$. Thus, the degree of accuracy of the final answer is no greater than that of the original data. If the answer had been rounded off to the second decimal place and stated as 11.91, the accuracy indicated would have been $0.005/11.91 \times 100 = 0.042\%$. This would indicate an accuracy greater than that of the original data.

An adequate degree of accuracy is maintained by using three *significant figures* and a 10-in. slide rule. All the following values are written to three significant figures: 8320, 8.32, and 0.0832. The location of the decimal does not affect the number of significant figures. These values and any others can be read to three significant figures on a 10-in. slide rule. Slide rule accuracy is between 0.1% and 1.0%. In many engineering areas, an accuracy of between 0.1% and 1.0% is adequate. For example, a calculation in stress analysis has to be based on at least two factors:

1. the strength of the material being used in the structure; and

2. the actual forces being applied to the structure.

Neither of these factors can be determined with a high degree of certainty. First, small as they might be, there are variations in both homogeneity of material and consistency of quality; second, the maximum load to which the structure will be subjected may have to be estimated.

For these reasons, calculations done with a 10-in. slide rule and to three significant figures are usually adequate. With the increasing popularity and use of the hand-held, electronic calculator, care must be taken not to indicate a misleading degree of accuracy.

1-5 MATHEMATICS REQUIRED

Although addition, subtraction, multiplication, and division are rather elementary, they are mentioned here to emphasize their importance in obtaining correct answers. Carelessness in such elementary mathematical operations is often the source of error as well as much frustration and wasted time in problem solutions. Knowledge of simple trigonometric functions and the sine and cosine laws is also necessary.

Trigonometry is the study of the relationships among the sides and interior angles of triangles. The angles are usually given Greek lowercase letters, some popular ones being alpha (α), beta (β), gamma (γ), theta (θ), and phi (ϕ).

The basic trigonometric functions apply only to right-angle triangles. The right-angle triangle can be in any one of four quadrants of an *x-y*-axis system (Figure 1-2). For each triangle, the functions are:

$$\text{sine}\,\theta = \sin\theta = \frac{\text{side opposite}}{\text{hypotenuse}}$$

$$\text{cosine}\,\theta = \cos\theta = \frac{\text{side adjacent}}{\text{hypotenuse}}$$

$$\text{tangent}\,\theta = \tan\theta = \frac{\text{side opposite}}{\text{side adjacent}}$$

The signs of these trig functions depend on the quadrant in which the triangle lies (Figure 1-2).

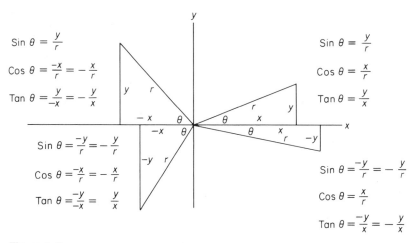

Figure 1-2

Example 1-2

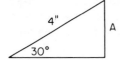

Figure 1-3a

Determine the value of length A for each triangle shown (Figure 1-3).

$$\sin 30° = \frac{A}{4}$$

$$A = \sin 30 \times 4$$
$$A = 0.5 \times 4$$
$$\underline{A = 2 \text{ in.}}$$

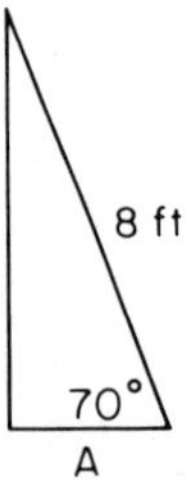

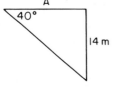

Figure 1-3b

$$\cos 70° = \frac{A}{8}$$

$$A = \cos 70 \times 8$$
$$A = 0.342 \times 8$$
$$\underline{A = 2.74 \text{ ft}}$$

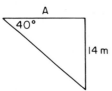

Figure 1-3c

$$\tan 40° = \frac{14}{A}$$

$$A = \frac{14}{\tan 40}$$
$$\underline{A = 16.7 \text{ m}}$$

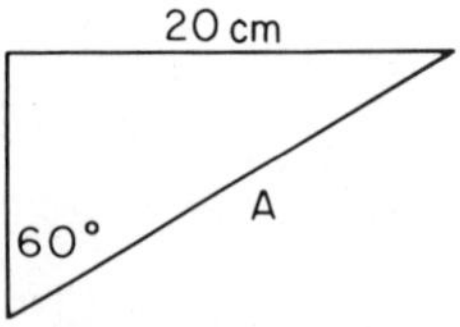

Figure 1-3d

$$\sin 60° = \frac{20}{A}$$

$$A = \frac{20}{\sin 60°}$$
$$A = \frac{20}{0.866}$$
$$\underline{A = 23.1 \text{ cm}}$$

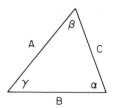

Figure 1-4

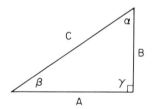

Figure 1-5

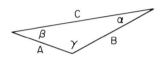

Figure 1-6

For triangles that are not right-angle triangles, such as the ones in Figures 1-4 and 1-5, either the sine law or the cosine law is used.

The sine law is given in Equation 1-1.

$$\frac{A}{\sin \alpha} = \frac{B}{\sin \beta} = \frac{C}{\sin \gamma} \qquad (1.1)$$

In each case, the side is divided by the sine of the angle opposite the side.

The cosine law is given in Equation 1-2.

$$C^2 = A^2 + B^2 - 2AB \cos \gamma \qquad (1-2)$$

If the cosine law is applied to a right-angle triangle where $\gamma = 90°$ (Figure 1-6) and $\cos \gamma = \cos 90° = 0$, the equation $C^2 = A^2 + B^2 - 2AB \cos \gamma$ becomes

$$C^2 = A^2 + B^2 \qquad (1-3)$$

Equation 1-3 is the *Pythagorean theorem*; it states that the square of the hypotenuse of a right-angle triangle equals the sum of the squares of the two remaining sides.

Example 1-3

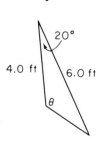

Figure 1-7

Find the length of the unknown side and the angle θ (Figure 1-7).

Designating this side C and using the cosine law gives:

$$C^2 = A^2 + B^2 - 2AB \cos \gamma$$
$$C^2 = (4)^2 + (6)^2 - 2 \times 4 \times 6 \times \cos 20°$$
$$C^2 = 16 + 36 - 48 \times 0.94$$
$$C^2 = 6.9$$
$$\underline{C = 2.6 \text{ ft}}$$

Using the sine law, we get:

$$\frac{6}{\sin\theta} = \frac{2.6}{\sin 20°}$$

$$\sin\theta = \frac{6 \times 0.342}{2.6}$$

$$= 0.789$$

$$\theta = 52.1°$$

But we know this to be in the second quadrant so

$$\theta = 180 - 52.1$$
$$\theta = 127.9°$$

Example 1-4

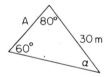

Figure 1-8

Find the length of side A (Figure 1-8).

The sum of all of the angles of a triangle is 180°. Therefore,

$$\text{angle } \alpha = 180° - (60 + 80) = 40°$$

Using the sine law, we obtain:

$$\frac{A}{\sin\alpha} = \frac{B}{\sin\beta}$$

$$\frac{A}{\sin 40°} = \frac{30}{\sin 60°}$$

$$A = 30 \times \frac{\sin 40°}{\sin 60°}$$

$$A = 30 \times \frac{0.643}{0.866}$$

$$A = 22.3 \text{ m}$$

Example 1-5

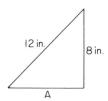

Figure 1-9

Find the length of side A (Figure 1-9).

$$(12)^2 = A^2 + (8)^2$$
$$A^2 = (12)^2 - (8)^2$$
$$= 144 - 64$$
$$= 80$$
$$\underline{A = 8.9 \text{ in.}}$$

PROBLEMS

1-1 A right-angle triangle has a hypotenuse of 30 in. and one angle of 40°. Determine the lengths of the other two sides.

1-2 A 15-m cord is fastened to the top of a flagpole and, when pulled taut, touches the ground 10 m from the flagpole. Find the height of the flagpole and the angle between it and the cord.

1-3 Calculate the angle θ and the hypotenuse R for each triangle shown in Figure P1-3.

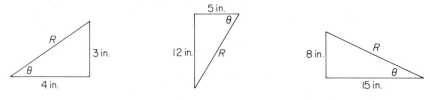

Figure P1-3a   **Figure P1-3b**   **Figure P1-3c**

1-4 For each triangle shown in Figure P1-4, determine the indicated unknown sides.

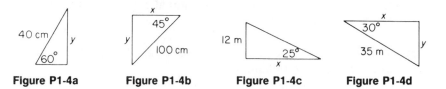

Figure P1-4a   **Figure P1-4b**   **Figure P1-4c**   **Figure P1-4d**

1-5 The top end of a 40-m conveyor can reach a height of 25 m. What is the angle between the conveyor and the ground?

1-6 A piece of sheet metal has a cross section as shown in Figure P1-6. Determine distance c.

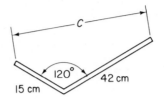

Figure P1-6

1-7 Determine the length of cylinder CB in Figure P1-7.

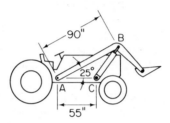

Figure P1-7

1-8 A surveying method to find the distance between two points A and B, between which is an obstacle, is shown in Figure P1-8. The length of CD is 640 ft. As an intermediate step of the surveying method, calculate the lengths of AC and AD.

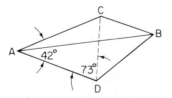

Figure P1-8

1-9 Determine the size of each of the internal angles of a triangle that has sides of 5, 13, and 16 cm.

1-10 A cone-shaped tube is 10 cm long and has diameters of 6 and 8 cm. Calculate the angle of taper.

1-11 A rotating shaft level gauge rotates from position A to B (Figure P1-11). What is the measured height of liquid?

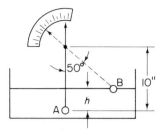

Figure P1-11

1-12 A part's dimensions may be determined in either of two ways as shown in Figure P1-12. Determine x and y.

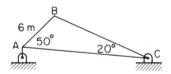

Figure P1-12a   **Figure P1-12b**

1-13 Determine the lengths of members AC and BC for the truss shown in Figure P1-13.

Figure P1-13

1-14 The width across the flats of a hexagonal nut is 1.875 in. What is the width across the corners?

1-15 Find the distance between any two holes of the plate shown in Figure P1-15.

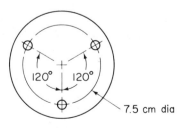

Figure P1-15

1-16 A roller and lever mechanism is in the position shown in Figure P1-16. Find the horizontal distance between A and B.

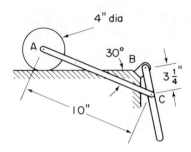

Figure P1-16

1-17 A sloped surface 15 m long is elevated at 25° to the horizontal. Determine the vertical rise and the horizontal run of the sloped surface.

1-18 Determine the length of side *d* of the triangle shown in Figure P1-18.

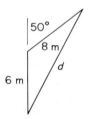

Figure P1-18

1-19 For the triangle in Figure P1-19, determine the length of side *d*.

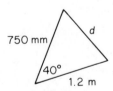

Figure P1-19

1-20   The top end of a 25-m pole is 18 m above the horizontal surface upon which the bottom end rests. What angle does the pole make with the horizontal surface?

1-21   The bottom end of a 6-m ladder is placed 2.5 m from point A. Determine: (a) distance d; and (b) the shortest distance from A to the ladder.

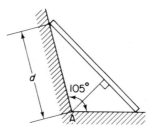

Figure P1-21

2

FORCES, VECTORS, AND RESULTANTS

2-1 VECTORS

Everyone feels that he knows what a force is and would probably define it as a push or pull. While this is true, there are further classifications. However, all forces do have one property in common: they are vector quantities and can be represented by vectors.

We must first distinguish between a *vector* quantity and a *scalar* quantity. You are no doubt familiar with scalar quantities. A board 10 ft long, a 2-hr time interval, a floor area of 20 m^2, and a 60-W light bulb all tell us "how much." These are scalar quantities; they indicate size or magnitude.

Vector quantities have the additional property of direction. Some vector quantities are a force of 15 N vertically downward, a distance of 20 km north, a velocity of 20 km/h east, and an acceleration of 7 ft/sec^2 upward. A vector quantity is, therefore, represented by an arrow; the arrowhead indicates the direction, and the length of the arrow indicates the magnitude.

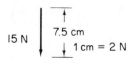

Figure 2-1

If we arbitrarily choose a scale of 1 cm = 2 N, a 15-N vertical force would be that shown in Figure 2-1. Similarly, the other vector quantities referred to previously would be those shown in Figures 2-2, 2-3, and 2-4.

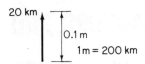

Figure 2-2

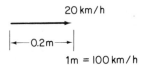

Figure 2-3

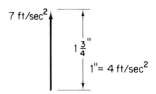

Figure 2-4

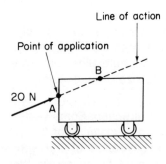

Figure 2-5

To be complete, both direction and magnitude must be labelled for each vector quantity. A vector representing a force has a point of application and a line of action. In Figure 2-5, a force of 20 N is applied to a cart. The vector shown indicates magnitude (20 N), a point of application (A), and the direction along the line of action.

If the 20 N force is not sufficient to move the cart, the cart has a balance of external forces acting on it and is said to be in *static equilibrium*.

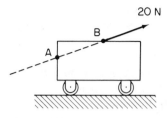

Figure 2-6

The principle of *transmissibility* states that a force acting on a body can be applied anywhere along the force's line of action without changing its effect on the body. Thus, the 20-N force can also be applied at point B as shown in Figure 2-6. Whether the point of application is A or B, the 20-N force will have the same effect on the cart. Later on in the text, this principle will be used quite frequently in dealing with the subject of *moments*.

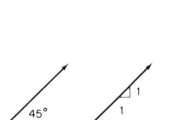

Figure 2-7a   **Figure 2-7b**

Vector quantities are not always vertical or horizontal. They may be at some angle or slope. The slope is indicated by reference to a horizontal line. This is done by giving either the angle in degrees, Figure 2-7(a), or the rise and run of the slope, Figure 2-7(b).

2-2   FORCE TYPES, CHARACTERISTICS, AND UNITS

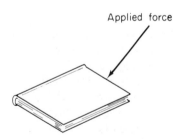

Figure 2-8

Realizing that forces can be represented by vectors, consider now the various types of forces. For the sake of easier discussion in a subject such as mechanics, several classifications are used. One such classification is that of *applied* and *nonapplied forces*. An applied force is a very real and noticeable force applied directly to an object. The force that you would apply to a book (Figure 2-8) to slide it across a table is an applied force. The nonapplied force acting on this same book (Figure 2-9) may not be as readily apparent since it is the force of gravity, or the weight of the book. Other examples of nonapplied forces are the force of magnetic attraction or repulsion and the force due to inertia.

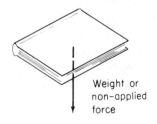

Figure 2-9

In analyzing various force systems later, keep in mind that forces such as weight and inertia are always present and may have to be included in your calculations.

Another classification categorizes forces as *internal* and *external*. The distinction here is very important since it is often the source of incorrectly drawn free-body diagrams (Section 4-2). Internal forces are often included where they should not be.

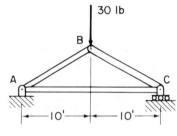

Figure 2-10

An internal force is a force inside a structure, and an external force is a force outside the structure. The pin-connected structure in Figure 2-10 has an external force of 30 lb. Knowing that connection C is on rollers and free to move horizontally, one can visualize that the horizontal member AC is *in tension*, i.e., there is a force tending to stretch it. The tensile force in AC is an internal force. There are also internal forces of compression in members AB and BC.

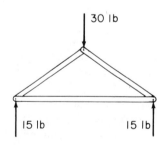

Figure 2-11

If we replace the actual supports at points A and C with equivalent supporting forces of 15 lb at each end (Figure 2-11), we then have a total of three external forces. These external forces can be further subdivided into *acting* and *reacting forces*. The 15-lb forces are present because of the 30-lb force; thus, they are a reaction to the application of the 30-lb force. Therefore, the 30-lb force is an *acting* force, and the 15-lb forces are *reacting* forces. You will be asked to solve for the reactions on various structures. *Reactions* are simply the reacting forces that are necessary to support the structure when its given method of support is removed.

As shown in Figure 2-11, the reaction at A is 15 lb vertically upward or $R_A = 15$ lb↑, and, similarly, $R_B = 15$ lb↑. (A quantity's being expressed in italic type will indicate that it is a vector quantity.)

Unless otherwise stated, the weight of all members or structures will be neglected to simplify problem solution. This can be done without appreciable error when the supported load is much greater than the structure's weight. When weights are significant in later problems, they will be included.

2-3   RESULTANTS

Scalar quantities such as 4 m² and 3 m² can be added to equal 7 m². But if we added vector quantities of 4 km and 3 km, their directions would have to be considered. This is known as *adding vectorially* or *vector addition*. The answer obtained is the *resultant*; it is a single vector giving the result of the addition of the original two or more vectors.

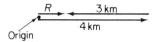

Figure 2-12a

What is the result of walking 4 km east and then 3 km west? You have walked a total distance of 7 km, but how far are you from your original location? Let each distance be represented by a vector [Figure 2-12(a)]. When adding 4 + 3 vectorially, place the vectors tip to tail. The resultant is a vector from the original point to the final point. In this case, $R = 1$ km in an easterly direction.

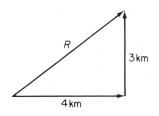

Figure 2-12b

Suppose that you had walked 4 km east and 3 km north. Again, the distances would be represented by vectors and added by being placed tip to tail [Figure 2-12(b)]. The resultant vector drawn from the original point to the final point is the hypotenuse of a right-angle triangle. For this case, $R = 5$ km in a northeast direction.

2-4   VECTOR ADDITION–GRAPHICAL

As was described in Section 2-3, the sum of two or more vectors is a resultant. The vector quantity used there was *distance*; other vector quantities often encountered are *force, velocity,* and *acceleration.* Force vectors are going to be our main concern in static mechanics.

Graphical vector addition requires the drawing of the vectors to some scale in their given direction. The resultant can then be measured or scaled from the drawing. This method of scale drawings will not be used to any extent in this text since drafting equipment is required and the accuracy of the solution is somewhat dependent on the scale used. However, graphical vector addition is still important, since the same drawings or sketches are required for the analytical method. Employing the analytical method, one uses the same sketches drawn roughly to scale and then solves mathematically.

The three methods or rules of vector addition are the *triangle, parallelogram,* and *vector polygon* methods. The triangle and parallelogram methods are essentially modifications of the vector polygon method. The triangle method of vector addition is shown in Figure 2-13.

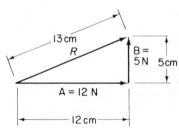

Constructing A and B [Figure 2-13(a)] to a scale of 1 cm=1 N, we can calculate the length of R to be 13 cm. R is therefore equal to 13 N.

Figure 2-13a

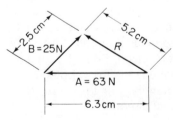

Using a scale of 1 cm=10 N [Figure 2-13(b)] and drawing B at the correct slope, we find the length of R to be 5.2 cm or R=52 N.

Figure 2-13b

Example 2-1

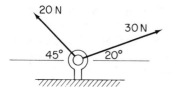

Figure 2-14a

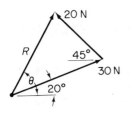

Figure 2-14b

Forces of 20 N and 30 N are pulling on a ring (Figure 2-14). Determine the resultant using the triangle rule.

Employing an appropriate scale such as 1 cm = 2 N and the angles of 20° and 45° as given, one can construct a vector triangle, Figure 2-14(b) or 2-14(c), giving an answer of $R = 28.2$ N $\nearrow 60°$. (Note that it is immaterial whether 20 N is added to 30 N or 30 N is added to 20 N.)

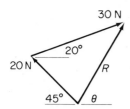

Figure 2-14c

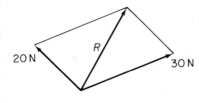

Figure 2-15

Applying the parallelogram rule to the forces in Example 2-1, we get the parallelogram shown in Figure 2-15. From the top of the 20-N force, a line is drawn parallel to the 30-N force. Similarly, a line is drawn parallel to the 20-N force. The diagonal of the parallelogram formed represents the resultant. The parallelogram is simply the two triangles of the triangle method.

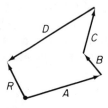

Figure 2-16

The vector polygon is a continuation of the triangle rule to accommodate more than two forces. Several vectors are added tip to tail—the sequence of the addition is not important. The resultant *R* is a vector from the origin of the polygon to the tip of the last vector (Figure 2-16).

Graphical vector addition is treated in more detail in the Appendix.

2-5 VECTOR ADDITION–ANALYTICAL

The solutions to mechanics problems are no exception to the old saying, "A picture is worth a thousand words." Sketches are vitally important in many cases; calculations should be accompanied by a sketch drawn as closely to scale as a little care will allow. A calculated answer can be visually checked for an obvious error in direction or magnitude.

Analytical vector addition consists of two main methods:

1. construction of a triangle and use of the cosine law or other simple trigonometric functions, or
2. addition of the components of vectors (Section 2-7).

Example 2-2

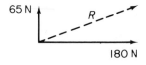

Figure 2-17a

Determine the resultant of the vectors shown in Figure 2-17(a). The resultant is easily obtained since the vectors, when added, form a right-angle triangle, as is shown in Figure 2-17(b).

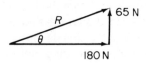

Figure 2-17b

$$R^2 = (180)^2 + (65)^2$$
$$R = \sqrt{32,400 + 4225}$$
$$= \sqrt{36,625}$$
$$R = 191 \text{ N}$$
$$\tan \theta = \frac{65}{180} = 0.361$$
$$\theta = 19.8°$$

The final answer is expressed as:

$$R = 191 \text{ N} \nearrow 19.8°$$

An alternate method would be

$$\tan \theta = 0.361$$
$$\theta = 19.8°$$
$$\sin \theta = \frac{65}{R}$$
$$R = \frac{65}{\sin 19.8°}$$
$$R = 191 \text{ N}$$

Note that, with this method, the value of R depends on a correct initial calculation of the value of θ. In the first solution, neither R nor θ was dependent on the other. Thus, the first method is preferred inasmuch as one mistake at the beginning does not make remaining calculations incorrect.

Example 2-3

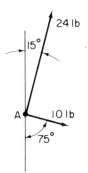

Figure 2-18a

Solve for the resultant of the force system shown in Figure 2-18 acting on point A.

$$R = \sqrt{(24)^2 + (10)^2}$$
$$R = 26 \text{ lb}$$
$$\tan \theta = \frac{10}{24} = 0.416$$
$$\theta = 22.6°$$

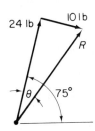

Figure 2-18b

Therefore:

$$75 - \theta = 75 - 22.6 \qquad [\text{Figure 2-18(b)}]$$
$$= 52.4°$$
$$R = 26 \text{ lb} \nearrow 52.4°$$

Example 2-4

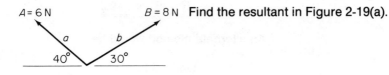

$A = 6$ N $B = 8$ N Find the resultant in Figure 2-19(a).

Figure 2-19a

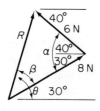

Figure 2-19b

Sketch a vector triangle [Figure 2-19(b)] and use the cosine law.

$$R^2 = A^2 + B^2 - 2AB\cos\alpha$$
$$R^2 = (6)^2 + (8)^2 - 2 \times 6 \times 8 \cos 70°$$
$$= 36 + 64 - 96 \times 0.342$$
$$= 100 - 32.8$$
$$R = \sqrt{67.2}$$
$$R = 8.2 \text{ N}$$

To show a direction for R, we must solve for $\theta = \beta + 30°$

By the sine law:

$$\frac{6}{\sin\beta} = \frac{8.2}{\sin 70°}$$

$$\sin\beta = \frac{6 \times 0.94}{8.2}$$

$$= 0.688$$

$$\beta = 43.5°$$
$$\theta = 43.5 + 30 = 73.5°$$

Our final answer is:

$$R = 8.2 \text{ N } \nearrow 73.5°$$

2-6 COMPONENTS

Previously, our main concern was the addition of two or more vectors to obtain a single vector, the resultant. *Resolution* of a vector is the reverse of the determination of the resultant.

A single force can be broken up into two separate forces. This is known as resolution of a force into its components. It is often convenient in problem solutions to be concerned only with forces in either the vertical or the horizontal direction. Therefore, the *x-y* axis system is used: a component in the horizontal direction has a subscript *x*, and a component in the vertical direction has a subscript *y*.

Example 2-5

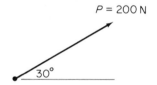

$P = 200$ N

Determine the horizontal and vertical components of P [Figure 2-20(a)].

30°

Figure 2-20a

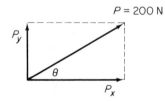

$P = 200$ N

Figure 2-20b

By constructing a parallelogram or rectangle, we are left with a right-angle triangle in which:

$$\sin \theta = \frac{P_y}{P}$$ [Figure 2-20(b)]

$$P_y = 200 \times \sin 30°$$
$$P_y = 200 \times 0.5$$
$$\underline{P_y = 100 \text{ N} \uparrow}$$

$$\cos \theta = \frac{P_x}{P}$$

$$P_x = 200 \times \cos 30°$$
$$P_x = 200 \times 0.866$$
$$\underline{P_x = 173 \text{ N} \rightarrow}$$

Example 2-6

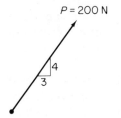

Figure 2-21a

Determine the horizontal and vertical components when the direction of P is shown as a slope [Figure 2-21(a)].

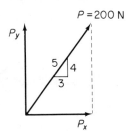

Figure 2-21b

Construct the vector triangle [Figure 2-21(b)]. Notice that we have two similar triangles. The hypotenuse of the small triangle $= \sqrt{(4)^2 + (3)^2} = 5$

Therefore:

$$\frac{4}{P_y} = \frac{5}{P} \quad \text{and} \quad \frac{3}{P_x} = \frac{5}{P}$$

or

$$P_y = \frac{4}{5} P \qquad\qquad P_x = \frac{3}{5} P$$

$$= \frac{4}{5} \times 200 \qquad\qquad = \frac{3}{5} \times 200$$

$$P_y = 160 \text{ N}\uparrow \qquad\qquad P_x = 120 \text{ N}\rightarrow$$

The components are not always in the horizontal and vertical directions, nor are they always at right angles to one another—as is illustrated in the following example.

Example 2-7

Find the components of force Q for the axis system of A and B as shown in Figure 2-22(a).

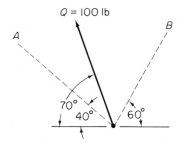

Figure 2-22a

Construct a vector parallelogram by drawing lines, parallel to axes A and B, from the tip of Q [Figure 2-22(b)]. Applying the sine law to the left half of the parallelogram, we obtain:

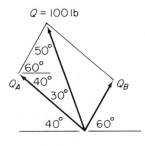

Figure 2-22b

$$\frac{Q_A}{\sin 50°} = \frac{Q}{\sin 100°}$$

$$Q_A = \frac{Q \sin 50°}{\sin 100°}$$

$$= \frac{Q \sin 50°}{\sin 80°}$$

$$= 100 \times \frac{0.766}{0.985}$$

$$Q_A = 77.8 \text{ lb } \underline{40° \nwarrow}$$

$$\frac{Q_B}{\sin 30°} = \frac{Q}{\sin 100°}$$

$$Q_B = \frac{Q \sin 30°}{\sin 80°}$$

$$= 100 \times \frac{0.5}{0.985}$$

$$Q_B = 50.8 \text{ lb } \underline{\nearrow 60°}$$

2-7 VECTOR ADDITION–COMPONENTS

In Section 2-4, a graphical solution was shown where several vectors were added by the use of a vector polygon. To add these vectors analytically using the method of components, one should proceed according to the following steps:

1. resolve each vector into a horizontal and vertical component;
2. add the vertical components, $R_y = \Sigma F_y$;

3. add the horizontal components, $R_x = \Sigma F_x$; and
4. combine the horizontal and vertical components to obtain a single resultant vector.

$$R = \sqrt{(R_x)^2 + (R_y)^2}$$

Note: The Greek letter Σ (sigma) means "the sum of." When writing $R_y = \Sigma F_y$, it is recommended that you say to yourself, "The resultant in the y-direction equals the sum of the forces in the y-direction."

Example 2-8

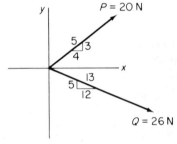

Figure 2-23

Find the resultant of forces *P* and *Q* as shown in Figure 2-23.

Using the x- and y-axes for the algebraic signs of the components, we have:

$$P_y = \frac{3}{5} \times 20 \qquad\qquad P_x = \frac{4}{5} \times 20$$

$$P_y = +12 \text{ N} \qquad\qquad P_x = +16 \text{ N}$$

$$Q_y = -\frac{5}{13} \times 26 \qquad\qquad Q_x = \frac{12}{13} \times 26$$

$$Q_y = -10 \text{ N} \qquad\qquad Q_x = +24 \text{ N}$$

$$R_y = 12 - 10 \qquad\qquad R_x = 16 + 24$$

$$R_y = 2 \text{ N} \qquad\qquad R_x = 40 \text{ N}$$

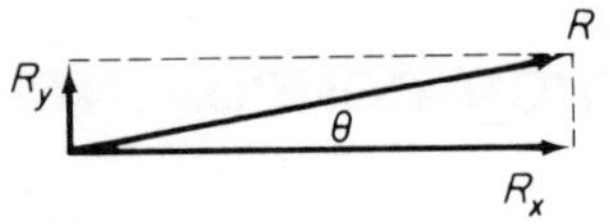

Figure 2-24

$$R = \sqrt{(2)^2 + (40)^2} \qquad \text{(Figure 2-24)}$$

$$R = 40.1 \text{ N}$$

$$\tan \theta = \frac{2}{40} = 0.05$$

$$\theta = 2.9°$$

$$\underline{R = 40.1 \text{ N} \; \nearrow 2.9°}$$

When determining the resultant of several forces, one may find it more convenient to tabulate all the components as follows:

FORCE	X	Y
P	+16 N	+12 N
Q	+24 N	−10 N
	$R_x = +40$ N	$R_y = +$ 2 N

Example 2-9

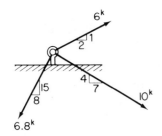

Figure 2-25

Determine the resultant of the forces shown in units of kips in Figure 2-25.

Figure 2-26a

For easier visualization, the components of each force can be drawn as in Figure 2-26.

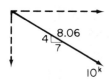

Figure 2-26b

Figure 2-26c

Tabulate the components as follows:

FORCE (IN KIPS)	X	Y
6^k	$\dfrac{2}{2.24} \times 6 = +5.35^k$	$\dfrac{1}{2.24} \times 6 = +2.68^k$
10^k	$\dfrac{7}{8.06} \times 10 = +8.69^k$	$\dfrac{4}{8.06} \times 10 = -4.97^k$
6.8^k	$\dfrac{8}{17} \times 6.8 = -3.2^k$	$\dfrac{15}{17} \times 6.8 = -6^k$
	$R_x = \overline{+10.84^k}$	$R_y = \overline{-8.29^k}$

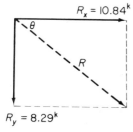

$R_x = 10.84^k$ From Figure 2-27

$$\tan\theta = \frac{8,29}{10.84}$$

$$\theta = 37.4°$$

$$R = \sqrt{(10.84)^2 + (8.29)^2}$$

$$R = 13.6^k \;\underline{\diagdown 37.4°}$$

$R_y = 8.29^k$

Figure 2-27

PROBLEMS

2-1 Force *P* is applied at point C as shown in Figure P2-1. Locate a point D at which *P* may be applied and still have the same effect upon the pins at A and B (as it does when applied at point C).

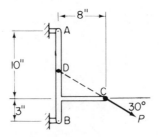

Figure P2-1

2-2 Locate distance *d* (Figure P2-2) such that the 8-kN force would have the same external effect on the frame if it were located at C rather than at B.

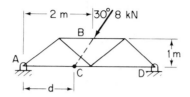

Figure P2-2

2-3 A 40-N force is applied to the lever at point B (Figure P2-3). It would have the same external effect if applied at point C or D. Locate these points with respect to pivot A by finding *x* and *y*. Determine the perpendicular distance between the force and pivot A.

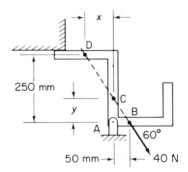

Figure P2-3

2-4 Determine the resultant and indicate the angle or slope for each force system in Figure P2-4.

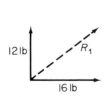

Figure P2-4a

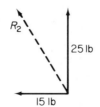

Figure P2-4b

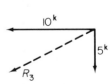

Figure P2-4c

2-5 Determine the resultant and indicate the angle or slope for each force system in Figure P2-5.

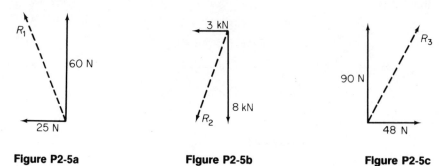

Figure P2-5a   **Figure P2-5b**   **Figure P2-5c**

2-6 A horizontal shaft exerts a sideways thrust of 200 N and a downward load of 600 N on a bearing. What is the resultant?

2-7 A ladder is supported at the floor by forces as shown in Figure P2-7. Determine the resultant of these forces.

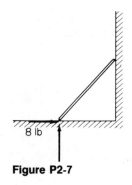

Figure P2-7

2-8 Find the resultant force acting on point A (Figure P2-8).

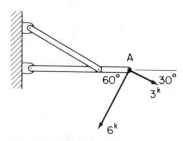

Figure P2-8

2-9 Member AB of the frame shown has two forces acting on it. Find the resultant force (Figure P2-9).

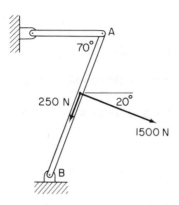

Figure P2-9

2-10 Determine the horizontal and vertical components of each force shown in Figure P2-10.

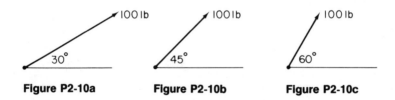

Figure P2-10a   **Figure P2-10b**   **Figure P2-10c**

2-11 Determine the horizontal and vertical components of each force shown in Figure P2-11.

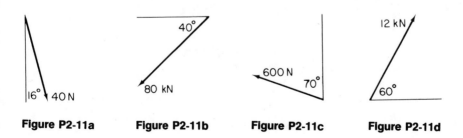

Figure P2-11a   **Figure P2-11b**   **Figure P2-11c**   **Figure P2-11d**

2-12 Determine the horizontal and vertical components of each force shown in Figure P2-12.

Figure P2-12a   Figure P2-12b   Figure P2-12c   Figure P2-12d   Figure P2-12e

2-13 Find the horizontal and vertical components of each vector in Figure P2-13.

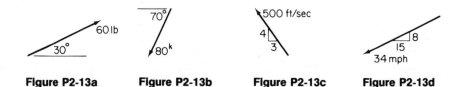

Figure P2-13a   Figure P2-13b   Figure P2-13c   Figure P2-13d

2-14 Find the horizontal and vertical components of each vector in Figure P2-14.

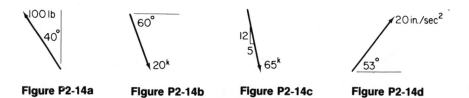

Figure P2-14a   Figure P2-14b   Figure P2-14c   Figure P2-14d

2-15 A crate is pulled up a slope by a force of 30 lbs. Find the component parallel to the slope and the component perpendicular to the slope (Figure P2-15).

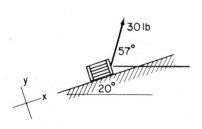

Figure P2-15

2-16   A purlin hanger is used to join timbers A and B (Figure P2-16). Timber B exerts a resultant force on the purlin of 1.8 kN $\searrow$80°. Calculate the horizontal and vertical components.

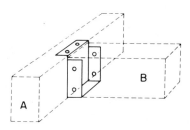

Figure P2-16

2-17   A frame supports a 65-kg block by means of a pulley and rope. Determine the horizontal and vertical components of the rope tension acting on A (convert kg to a force in Newtons). (See Figure P2-17.)

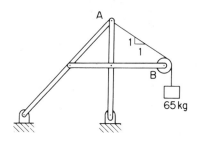

Figure P2-17

2-18   Resolve each force shown in Figure P2-18 into components in the x- and y-direction for the given x-y-axes orientation [Figure P2-18(a)].

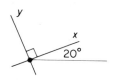

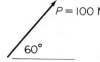

Figure P2-18a   **Figure P2-18b**   **Figure P2-18c**   **Figure P2-18d**

2-19 A lawnmower is pushed up a 15° slope by a 20-lb force. Find the component acting parallel to the slope (Figure P2-19).

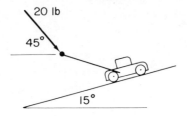

Figure P2-19

2-20 Find the resultant of the forces shown in Figure P2-20.

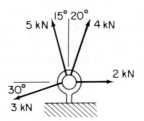

Figure P2-20

2-21 Find the resultant of the forces shown in Figure P2-21.

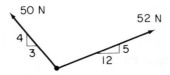

Figure P2-21

2-22 A cantilever beam supports forces as shown in Figure P2-22. Determine the resultant.

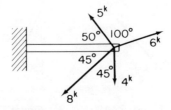

Figure P2-22

2-23 Find the resultant of the force system shown in Figure P2-23.

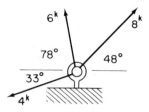

Figure P2-23

2-24 Find the resultant of the force system shown in Figure P2-24.

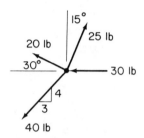

Figure P2-24

2-25 Find the resultant of the force system shown in Figure P2-25.

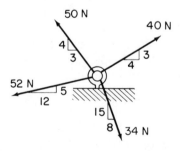

Figure P2-25

2-26 Find the resultant of the force system shown in Figure P2-26.

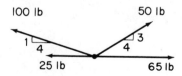

Figure P2-26

2-27   A carton is pushed onto a platform by the forces shown in Figure P2-27. Find the resultant force.

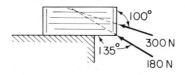

Figure P2-27

2-28   Using the cosine law, solve for the resultant of the forces shown in Figure P2-28.

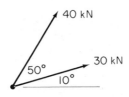

Figure P2-28

2-29   A telephone line forms an angle of 130° when it changes direction at a pole. The horizontal component of the tension in the telephone line is 75 lb. Find the resultant horizontal force on the telephone pole.

2-30   One method of clearing bush from land is to use a large steel ball with chains attached pulled by two caterpillar tractors (Figure P2-30). The chain tensions are $AB=4$ kips and $CD=3$ kips. Find the resultant force on the steel ball.

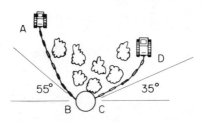

Figure P2-30

3

MOMENTS AND
COUPLES

3-1 MOMENT OF A FORCE

Moment is merely another term meaning *torque*, which is something producing or tending to produce rotation or torsion. Common examples of moment or torque are numerous. Pushing on a revolving door while walking through it, tightening a nut with a wrench, and turning the steering wheel of a car all involve moments. The moment is present whether there is actual rotation or only a tendency to rotate. A moment is caused by some force acting at a perpendicular distance from an *axis* (or *pivot point*):

moment = force × the perpendicular
distance between the axis and the
line of action of the force

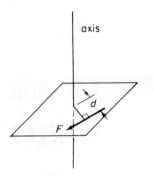

Figure 3-1

In Figure 3-1, the moment, $M = F \times d$, is in a clockwise direction. The units of moment are pound-feet (lb-ft), pound-inches (lb-in.), or kip-feet (kip-ft). When expressed this way, moment is easily distinguished from work, which is stated in units of ft-lb and in.-lb. In the SI metric system, force is expressed in newtons, and distance in meters. The units of moment or torque are newton-meter (N·m). Recall that a newton is the force necessary to give a mass of 1 kg an acceleration of 1 meter/s/s. A mass of 1 kg exposed to the standard acceleration of gravity ($g = 9.81$ m/s^2) will weigh 9.81 N. Keep in mind that the term kilogram is reserved for units of mass and should not be used as a unit of force even though it was used for force in the old MKS metric system.

In two-dimensional drawings—which are frequently used—the axis appears as a point. Moments taken about a point are indicated as being clockwise ($\curvearrowright$) or counterclockwise ($\curvearrowleft$). For the sake of uniformity in calculations, we will assume clockwise to be negative and counterclockwise to be positive.

Moments are vector quantities, and their direction must be indicated in one of three ways. For example, the same moment can be expressed as 10 lb-ft $\curvearrowright$, -10 lb-ft, or 10 lb-ft clockwise.

Example 3-1

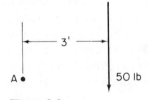

Figure 3-2

Calculate the moment about point A, Figure 3-2.

$$M = F \times d$$
$$= -(50 \times 3)$$
$$\underline{M = 150 \text{ lb-ft } \curvearrowright}$$

Example 3-2

Calculate the moment about point B due to

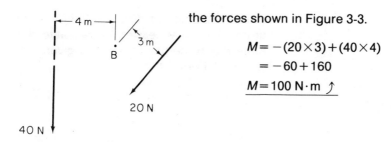

the forces shown in Figure 3-3.

$$M = -(20 \times 3) + (40 \times 4)$$
$$= -60 + 160$$
$$M = 100 \text{ N·m} \;\curvearrowleft$$

Figure 3-3

It is frequently easier to calculate moments by using a distance and a perpendicular force instead of a force and a perpendicular distance.

Example 3-3 is solved by two methods, the second method illustrates the breaking down of a 100-lb force into its components. The total moment is equal to the sum of each component times its perpendicular distance.

Example 3-3

Solve for the moment about A due to the 100-lb force (Figure 3-4).

Method (1)

moment = force × perpendicular distance
$$= -(100 \times 8)$$
$$M = 800 \text{ lb-in} \;\curvearrowright$$

Figure 3-4

Method (2)

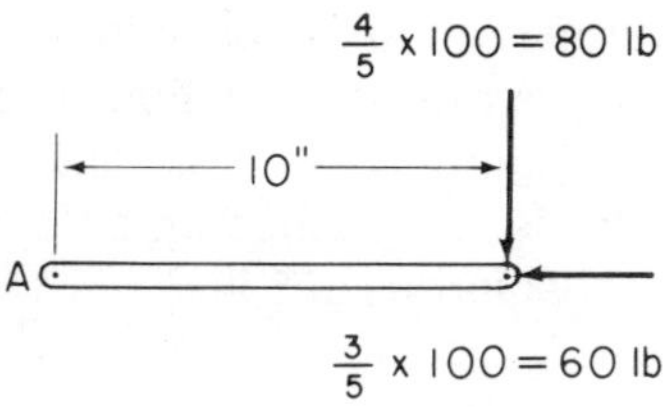

Resolve the 100-lb force into horizontal and vertical components (Figure 3-5).

$$M = -(80 \times 10) + (60 \times 0)$$
$$= -800 + 0$$
$$M = 800 \text{ lb-in} \;\curvearrowright$$

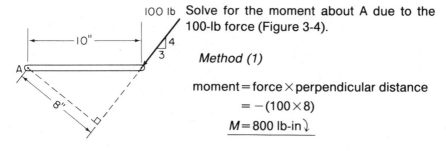

Figure 3-5

If the perpendicular distance (8 in.) is not given, it is often easier to calculate the force (80 lb) perpendicular to a given distance (10 in.).

Example 3-4

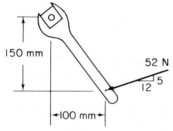

Figure 3-6a

Calculate the moment about the center of the nut [Figure 3-6(a)].

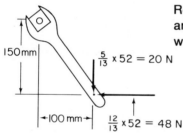

Figure 3-6b

Resolving the 52 N force into horizontal and vertical components [Figure 3-6(b)], we have:

$$M = F_1 \times d_1 + F_2 \times d_2$$
$$= -(20 \times 100) - (48 \times 150)$$
$$= -2000 - 7200$$
$$= 9200 \; N \cdot mm$$
$$\underline{M = 9.2 \; N \cdot m \; \curvearrowright}$$

3-2 COUPLES

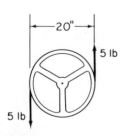

Figure 3-7

Let us look further at an application of moments by considering the rotation of a steering wheel (Figure 3-7). It is pulled down on the left side and pushed up on the right side with equal forces of 5 lb.

The moment due to the right-hand force is 5×10, or 50 lb-in. ⟳ .

$$\text{total moment} = 50 + 50$$
$$= 100 \text{ lb-in } ⟳$$

These forces could have been treated as a *couple*, that consists of two forces that are:

1. equal;

2. acting in opposite directions; and

3. separated by some perpendicular distance d.

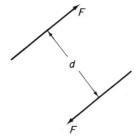

Figure 3-8

These three requirements of a couple are shown in Figure 3-8. Referring back to the steering wheel in Figure 3-7, we have

$$\text{couple moment} = F \times d$$
$$= 5 \times 20$$
$$\underline{M = 100 \text{ lb-in } ⟳}$$

This is the same answer that we obtained when we multiplied the individual forces by their distances from the pivot.

To illustrate how easily you can be mistaken and assume that two forces are a couple when they really are *not*—all the systems of Figure 3-9 are *not* couples for the following reasons:

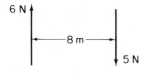

Figure 3-9a

a. The forces are not equal.

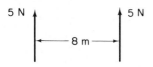

Figure 3-9b

b. The forces are not in opposite directions.

Figure 3-9c

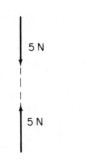

Figure 3-9d

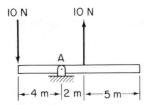

Figure 3-10a

c. The forces are neither parallel nor in opposite directions.

d. The forces are not separated by a distance *d.*

You will have noticed that when we calculated moments (Section 3-1), we specified the axis or pivot point about which the moments were calculated. It does not matter where the pivot point is located when we deal with couples. A couple has the same moment about all points. To illustrate this, consider a lever [Figure 3-10(a)] loaded as shown with its pivot point located at A.

Neglecting the weight of the lever and considering the forces as a couple, we can calculate the moment of the couple:

$$10 \times 6 = 60 \text{ N·m } \circlearrowleft$$

Checking this value by considering the moment of each force about A, we have:

$$M = (10 \times 4) + (10 \times 2)$$
$$= 40 + 20$$
$$\underline{M = 60 \text{ N·m } \circlearrowleft}$$

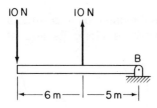

Figure 3-10b

Taking the same lever and moving the pivot to B [Figure 3-10(b)], we now have:

$$M = (10 \times 11) - (10 \times 5)$$
$$= 110 - 50$$
$$M = 60 \text{ N·m } \circlearrowleft$$

Thus, regardless of the pivot location, we still have a couple of 60 N·m in a counter-clockwise direction.

Suppose now that we have a fixed pivot and move the couple. In each part of Figure 3-11, a couple of 60 N·m is located on the lever.

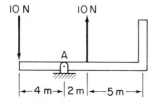

Figure 3-11a

The moment calculation for each is as follows:

a. $M = (10 \times 4) + (10 \times 2)$
$= 40 + 20$
$M = 60 \text{ N·m } \circlearrowleft$

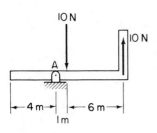

Figure 3-11b

b. $M = -(10 \times 1) + (10 \times 7)$
$= -10 + 70$
$M = 60 \text{ N·m } \circlearrowleft$

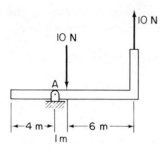

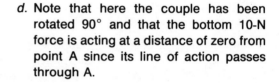

Figure 3-11c

c. The moment is the same as in *b* since a force acts anywhere along its line of action.

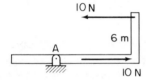

Figure 3-11d

d. Note that here the couple has been rotated 90° and that the bottom 10-N force is acting at a distance of zero from point A since its line of action passes through A.

$$M = (10 \times 0) + (10 \times 6)$$
$$M = 60 \ N \cdot m \ \smallint$$

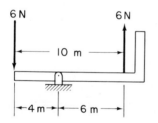

Figure 3-11e

e. The forces and perpendicular distance have changed, but the couple moment is the same
$(6 \times 10 = 60 \ N \cdot m)$.

$$M = (4 \times 6) + (6 \times 6)$$
$$= 24 + 36$$
$$M = 60 \ N \cdot m \ \smallint$$

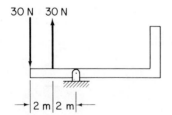

Figure 3-11f

f. $M = (30 \times 4) - (30 \times 2)$
$$= 120 - 60$$
$$M = 60 \ N \cdot m \ \smallint$$

The highlights to remember about couples are:

1. a couple is always characterized by the three properties listed previously;
2. the couple moment is unaffected by the pivot location; and
3. a couple can be shifted and still have the same moment about a given point.

Example 3-5

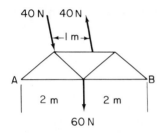

Figure 3-12

A *truss* is a structure composed of bars or members connected together to form one or more connected triangles. Calculate the moment about point A of the truss shown in Figure 3-12.

moment about A = moment of a couple
                    + moment of 60-N force

$$M_A = (40 \times 1) - (60 \times 2)$$
$$= 40 - 120$$
$$\underline{M_A = 80 \ \text{N} \cdot \text{m} \ \natural}$$

Example 3-6

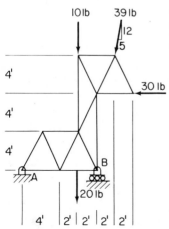

Figure 3-13a

Calculate the moment about point A due to the forces on the truss shown in Figure 3-13(a).

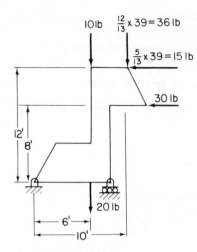

Figure 3-13b

Since we are concerned only with the complete truss and the external forces shown, we can break the 39-lb force into horizontal and vertical components; a structure such as the one in Figure 3-13(b) will be the result.

Taking moments about point A, we get:

$$M = -(10 \times 6) - (20 \times 6) - (36 \times 10)$$
$$+ (15 \times 12) + (30 \times 8)$$
$$= -60 - 120 - 360 + 180 + 240$$
$$= -540 + 420$$
$$M = 120 \text{ lb-ft} \;\curvearrowright$$

PROBLEMS

3-1 Calculate the moment about point A (Figure P3-1). (Each section represents 1 ft².)

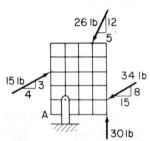

Figure P3-1

3-2 Calculate the moment about point A in Figure P3-2. (Each section represents a 1 m².)

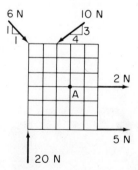

Figure P3-2

3-3 Figure P3-3 shows the view from above of the forces that workmen must apply to move a crate on the floor. What is the moment about the center of the crate?

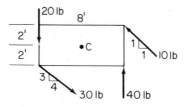

Figure P3-3

3-4 In an effort to tip a crate about the edge shown as point A (Figure P3-4), three forces are applied. Determine the moment about A due to these forces.

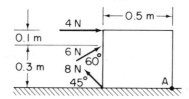

Figure P3-4

3-5 Calculate the moment about point A in Figure P3-5.

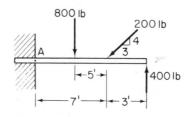

Figure P3-5

3-6 The bending moment in the beam at point A is equal to the moment of the 800-N force about point A. Calculate the moment about point A (Figure P3-6).

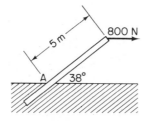

Figure P3-6

3-7 Calculate the moment about point A in Figure P3-7.

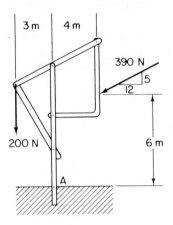

Figure P3-7

3-8 Calculate the moment about point A due to the 150-N force shown in Figure P3-8.

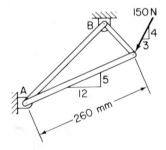

Figure P3-8

3-9 Calculate the moment or torque tightening the pipe in Figure P3-9.

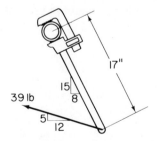

Figure P3-9

3-10 Find the moments about pins A and B due to the forces shown in Figure P3-10.

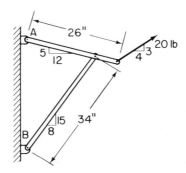

Figure P3-10

3-11 Find the moment about pins A and B due to the forces shown in Figure P3-11.

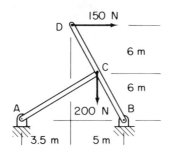

Figure P3-11

3-12 Find the moment about point A due to the forces shown in Figure P3-12.

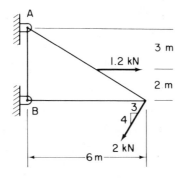

Figure P3-12

3-13 to 3-17   The forces shown form a couple. Replace the shown couple with an equivalent couple acting at points A and B.

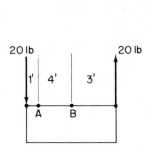

Figure P3-13

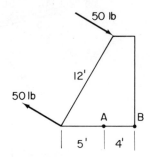

Figure P3-14

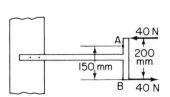

Figure P3-15

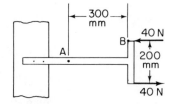

Figure P3-16

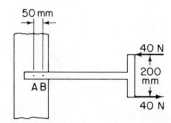

Figure P3-17

3-18 Find the moments about points A and B for the forces shown in Figure P3-18.

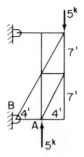

Figure P3-18

3-19 Find the moments about points A and B for the truss shown in Figure P3-19.

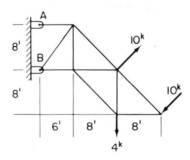

Figure P3-19

3-20 Find the moment about point A for the forces shown in Figure P3-20.

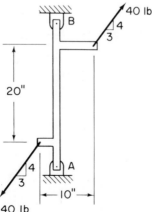

Figure P3-20

3-21 Find the moments about points A and B due to the forces shown in Figure P3-21.

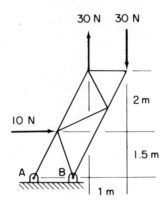

Figure P3-21

EQUILIBRIUM

4-1 THREE EQUATIONS OF EQUILIBRIUM

Initial understanding of this section will be easier if we consider only coplanar force systems. *Coplanar forces* are those that act only in one plane, such as the sheet of paper that they are drawn on. All the force systems that we have used so far have been coplanar.

For the remainder of the statics portion of this text, we will be considering objects said to be in *static equilibrium*. Static means "at rest," and equilibrium means that all forces applied on the object are in balance or have an equal and opposite force. The chair in which you sit is in static equilibrium. It is not moving up or down; therefore, total forces up equal total forces down. In this case, your weight and that of the chair are equal to the force of the floor pushing up on the chair.

The sideways forces are also equal. An externally applied sideways force that fails to move the chair is equal to the force of friction that the floor exerts on the chair.

A stationary chair is not twisting or turning either. This is because any clockwise (cw) moment has an equal and opposite counterclockwise (ccw) moment.

If the chair begins to move, it does so because there is an imbalance of forces on it, and static equilibrium no longer exists.

For complete static equilibrium, three requirements must be met:

1. vertical forces balance;
2. horizontal forces balance;
3. moments balance cw = ccw (about any point).

A concise form of stating the same points is:

1. $\Sigma F_y = 0$
2. $\Sigma F_x = 0$
3. $\Sigma M = 0$

If we take upward vertical forces as positive and downward vertical forces as negative, then the algebraic sum of all vertical forces is zero and a body is in balance or static equilibrium.

$\Sigma F_y = 0$, stated in words, is "summation of forces in the y-direction equals zero," or "forces up minus forces down equals zero."

If horizontal forces are positive to the right and negative to the left, then we can also say, "summation of forces in the *x-direction equals zero,*" or *"forces to the right minus forces to the left equals zero"* ($\Sigma F_x = 0$).

The same principle applies to moments. The sum of counterclockwise moments equals the sum of clockwise moments about *any* point on the object ($\Sigma M = 0$). Taking clockwise moments as negative and counterclockwise as positive, we find that the algebraic sum of moments is zero.

4-2   FREE-BODY DIAGRAMS

The next step is to draw an object and to show all the forces acting on it so that any or all of the three equilibrium equations may be applied.

$$\Sigma F_y = 0$$
$$\Sigma F_x = 0$$
$$\Sigma M = 0$$

To do this, we isolate the object as if it were floating freely in space and call it a *free-body diagram* (FBD). The object appears to be floating in space because we have removed all visible support or physical contact that it previously had. We replace these supports or physical contacts with equivalent forces, which have the same effect as the initial supports had. A free-body diagram of an object thus shows the forces acting *on* the object.

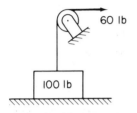

Figure 4-1

The crate in Figure 4-1 is drawn as a free-body diagram in Figure 4-2 and is acted upon by the following forces:

1. The rope tension is replaced by a force of 60 lb pulling upward on the crate.
2. The pull of gravity is 100 lb on the crate.
3. The floor had been partially supporting or pushing up on the crate. The support of the floor on the crate is shown as *N*.

Free−Body Diagram of Crate

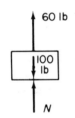

Figure 4-2

A final check is made to ensure that no forces have been omitted and that our free-body diagram is complete. Starting calculations with a partially completed free-body diagram can only lead to errors.

The equation to be used for Figure 4-2 would be

$$\Sigma F_y = 0$$

or

$$\text{forces up} - \text{forces down} = 0$$
$$N + 60 - 100 = 0$$
$$N = +40$$
$$\underline{N = 40 \text{ lb}\uparrow}$$

If, when you are drawing a free-body diagram, you are confused as to what forces are acting on the object and in which direction each is acting, visualize yourself in the place of the object. Ask yourself the following questions: What forces would be acting on me? Where would they be acting? Would they be pushing or pulling? The forces are of three categories:

1. the applied forces;
2. the nonapplied forces such as weight; and
3. the forces replacing a support or sectioned member.

A free-body diagram of a member is thus a picture showing how the rest of the world is acting *on* the member, not what the member is doing *to* anything else.

4-3   FREE-BODY DIAGRAM CONVENTIONS

When drawing free-body diagrams and replacing supports with equivalent supporting forces, one must employ some definite assumptions or conventions. Figure 4-3 shows the forces drawn to replace various supports or connections on the main member.

a. roller—the roller cannot exert a horizontal force; therefore, only a force perpendicular to the surface is present.

Figure 4-3a

b. a second method of representing a roller—the only force present is that perpendicular to the roller surface.

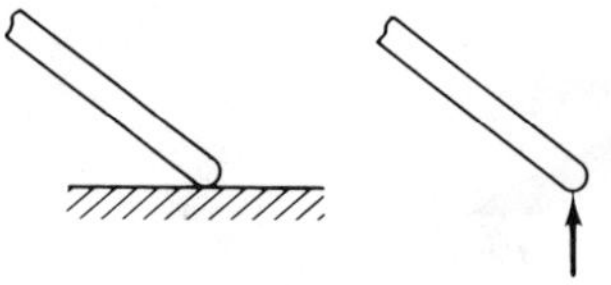

Figure 4-3b

c. smooth surface—zero friction is assumed; therefore, only one force, that perpendicular to the surface, is present.

Figure 4-3c

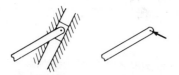

d. a *slot* uses the same principle as a smooth surface—there is only one force present, that perpendicular to the slot.

Figure 4-3d

Figure 4-3e

e. pinned—both horizontal and vertical components must be assumed at a pinned connection unless it is on a roller or smooth surface.

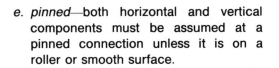

Figure 4-3f

f. the orientation of the support is immaterial—a horizontal and *vertical* force must still be assumed.

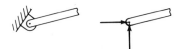

Figure 4-3g

g. cable—there is always a single force pulling in the direction of the cable.

As indicated earlier (Section 2-2), all objects or members will be assumed to be weightless unless the weight is specifically stated. When considering weight later on, one can easily include it in the calculations by adding another force passing through the center of gravity of the object or member.

Another assumption is that of zero friction on smooth surfaces. This simplifies calculations for our initial problems. When friction is to be included later, the coefficient of friction or some other clear indication that it must be considered will be given.

Since complete free-body diagrams are so crucial to correct problem solution, the following examples show free-body diagrams of various members or structures.

If you experience some difficulty in deciding whether vertical force is up or down, do not worry. If you assume an incorrect direction, the calculated answer will be negative, and you just change the direction of your original vector; all calculations are unchanged. If this answer must be used in further calculations, substitute it into equations as a negative value and do not change any vector directions until the calculations are complete.

When using the free-body diagram conventions, note that the horizontal bar of Figure 4-4 has horizontal and vertical forces at B but only a vertical force at A.

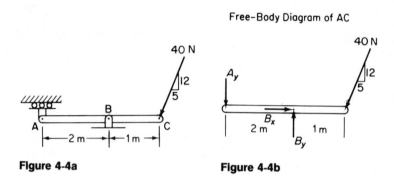

Figure 4-4a   **Figure 4-4b**

In Figure 4-5, a pin fastened to BE is free to slide in the slot of AC. Note the difference in the direction of B in the free-body diagrams of BE and AC. The slot of AC acts downward to the right *on BE*, and BE acts upward to the left *on AC*. This is the case of equal and opposite forces, with their direction depending on the object for which a free-body diagram is drawn. Note that force B has a slope of $\frac{3}{4}$ and is perpendicular to member AC, which has a slope of $\frac{4}{3}$. This illustrates that whenever you need the slope of a line that is perpendicular to another line, you simply reverse the slope numbers.

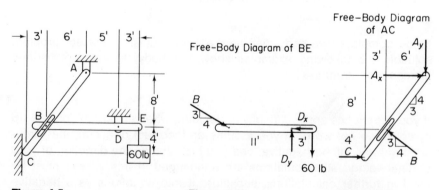

Figure 4-5a   **Figure 4-5b**   **Figure 4-5c**

Since cables can only be in tension (not compression), the free-body diagram of ring B (Figure 4-6) shows three forces pulling on B.

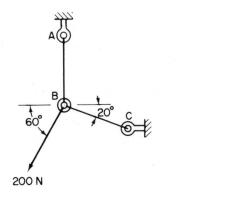

Figure 4-6a

Free-Body Diagram of B

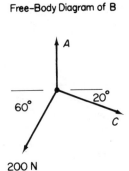

Figure 4-6b

Note that while there may be loads in each member of the pin-connected structure shown in Figure 4-7, they are internal loads and are thus of no concern in a free-body diagram of the complete frame; only external forces must be accounted for.

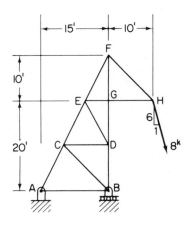

Figure 4-7a

Free-Body Diagram of Frame

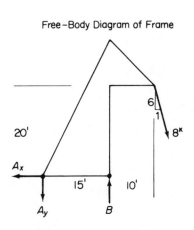

Figure 4-7b

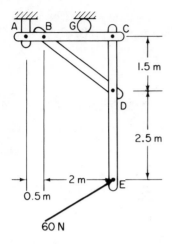

Figure 4-8a

From the frame of Figure 4-8(a), three free-body diagrams are drawn. In the free-body diagram of BD [Figure 4-8(b)], it is not necessary to show horizontal and vertical components at pinned connections B and D. Since member BD is in tension only, the total force at B must have the same direction or line of action as member BD. If the line of action of force *B* had not coincided with member BD, an unbalanced moment would have been present, and we would not have had static equilibrium. This can be proven by applying two of the basic equilibrium equations and finding $B_x = D_x$ and $B_y = D_y$.

Free–Body Diagram of BD

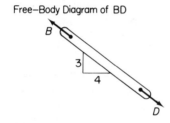

Figure 4-8b

Notice how the direction of the forces at B, C, and D changes depending on the member for which the free-body diagram is being drawn.

Free–Body Diagram of CE

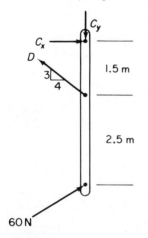

Figure 4-8c

Free–Body Diagram of AC

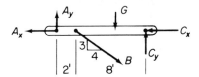

Figure 4-8d

4-4 TWO-FORCE MEMBERS

A member that is acted upon by two forces—for example, one at each end—is known as a *two-force member*. A two-force member will always be in either tension or compression.

When a member is acted upon by at least three forces at several locations, there will be not only compression or tension but also bending. In Figure 4-8(a), member BD is a two-force member, and members AC and CE are three-force members.

As was stated earlier, when drawing a free-body diagram of a pinned connection, one must assume both horizontal and vertical components. This rule still applies to two-force members, which are in tension or compression. In the case of the two-force member, the direction of the resultant of the horizontal and vertical components is known. This resultant must be directly in line with the member.

Free–Body Diagram of BD

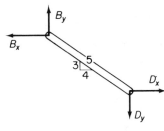

Figure 4-9

The free-body diagram of BD can be drawn as in either Figure 4-9 or Figure 4-10. For BD to be in static equilibrium and not to rotate, forces B and D must be equal, opposite, and must have the same line of action as the member.

$$B_y = D_y$$
$$B_x = D_x$$

and

$$B = B_x + B_y \quad \text{(vectorially)}$$
$$D = D_x + D_y \quad \text{(vectorially)}$$

Free-Body Diagram of BD

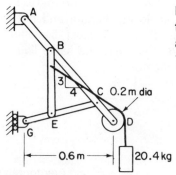

Figure 4-10

Suppose that $B_x = 80$ N. Knowing the slope of the member and by similar triangles, we get:

$$\frac{B_y}{B_x} = \frac{3}{4}$$

$$B_y = \frac{3}{4} \times B_x$$

$$= \frac{3}{4} \times 80$$

$$B_y = 60 \text{ N}$$

Similarly,

$$B = \frac{5}{4} \times B_x$$

$$= \frac{5}{4} \times 80$$

$$\underline{B = 100 \text{ N}}$$

Thus, for two-force members, the one component can be used to solve for the other component.

4-5 PULLEYS

Free-body diagrams of pulleys or structures that have pulleys may require careful analysis. Consider the pulley in Figure 4-11.

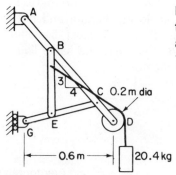

Figure 4-11

Free-Body Diagram of Pulley

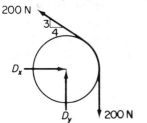

Figure 4-12

Cable tension due to the 20.4 kg mass is $20.4 \times 9.81 = 200$ N. The free-body diagram of the pulley would have two forces of 200 N and horizontal and vertical components at D (Figure 4-12).

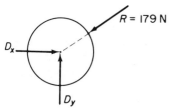

Figure 4-13

Adding the 200-N forces vectorially, we obtain a resultant, $R = 179$ N. For moment equilibrium about the center of the pulley, the resultant R must pass through the center (Figure 4-13).

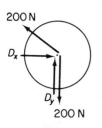

Figure 4-14

Since R can be applied anywhere along its line of action, let us suppose that it acts at D and resolve it into its original components, i.e., two 200-N forces (Figure 4-14).

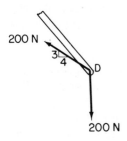

Figure 4-15

In a free-body diagram of member AD, the forces at D would be those shown in Figure 4-15. Regardless of pulley diameter, the rope tensions acting on the pulley can be shown as acting at the center of the pulley. A free-body diagram of the complete structure would be either the one in Figure 4-16 or the one in Figure 4-17.

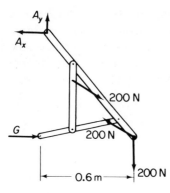

In Figure 4-16, there are three 200-N external forces with the vertical one acting 0.6 m from point G. In Figure 4-17, there is only one 200-N external force (the remaining rope tension is an internal force). Notice, however, that the 200-N vertical force is now acting 0.7 m from point G. In this case, the diameter of the pulley is significant.

Figure 4-16

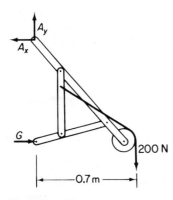

Either figure is correct, and the solution of either will yield the same answers for A_y, A_x, and G. (Figure 4-17 would be easier to solve.) Remember to analyze each problem individually—do not blindly ignore pulley diameters.

Figure 4-17

4-6 COPLANAR CONCURRENT FORCE SYSTEMS

With a *coplanar concurrent* force system, we usually have a free-body diagram of a point and can only apply the two equations of equilibrium, $\Sigma F_x = 0$ and $\Sigma F_y = 0$. Two unknowns and simultaneous equations may result. An alternate solution consists of drawing a vector polygon. If there are only three forces, the result is a vector triangle to which the sine law can be applied.

In Section 2-4, the vector triangle was used to obtain a resultant. The resultant was the third and unknown force required to complete the triangle; it began at the origin and was directed away from the origin. For a system to be in equilibrium, the equilibrant force must be equal and opposite to the resultant of all other forces. The equilibrant force vector then closes the triangle or polygon as the resultant force vector did, but it points *toward* the origin rather than *away* from it.

Example 4-1

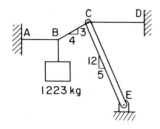

Figure 4-18

Find the load in each section of the cable system in Figure 4-18.

$$1223 \text{ kg} \times 9.81 \text{ m/s}^2 = 12 \text{ kN}$$

Free-Body Diagram of B

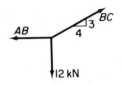

Figure 4-19

Starting with a free body diagram that has a known force (Figure 4-19), we have:

$$\Sigma F_y = 0$$

$$\frac{3}{5} BC - 12 = 0$$

$$BC = \frac{5 \times 12}{3}$$

$$\underline{BC = 20 \text{ kN } T}$$

(T will be used to represent tension; C will represent compression.)

$$\Sigma F_x = 0$$

$$-AB + \frac{4}{5} \times 20 = 0$$

$$\underline{AB = 16 \text{ kN } T}$$

Free-Body Diagram of C

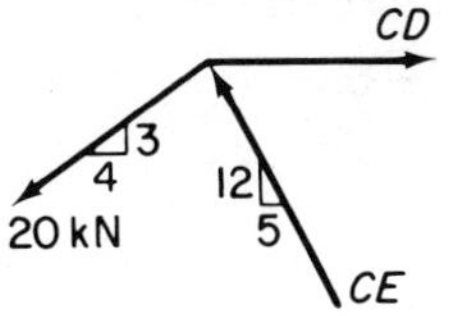

CD

3
4
20 kN
12
5
CE

Figure 4-20

$$\Sigma F_y = 0 \qquad\qquad \text{(Figure 4-20)}$$

$$\frac{12}{13} CE - \frac{3}{5} \times 20 = 0$$

$$CE = \frac{12 \times 13}{12}$$

$$\underline{CE = 13 \text{ kN } C}$$

$$\Sigma F_x = 0$$

$$CD - \frac{4}{5} \times 20 - \frac{5}{13} \times 13 = 0$$

$$CD = 16 + 5$$

$$\underline{CD = 21 \text{ kN } T}$$

Example 4-2

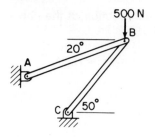

Figure 4-21

Determine the loads in members AB and CB of Figure 4-21.

Method 1—Simultaneous Equations

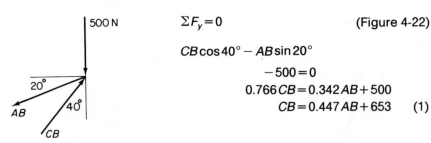

Figure 4-22

$$\Sigma F_y = 0 \qquad \text{(Figure 4-22)}$$

$$CB\cos 40° - AB\sin 20°$$
$$-500 = 0$$
$$0.766\,CB = 0.342\,AB + 500$$
$$CB = 0.447\,AB + 653 \qquad (1)$$

$$\Sigma F_x = 0$$
$$- AB\cos 20° + CB\sin 40° = 0$$
$$AB \times 0.94 = CB \times 0.643$$
$$AB = 0.684\,CB \qquad (2)$$

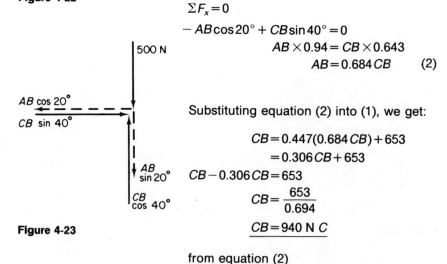

Figure 4-23

Substituting equation (2) into (1), we get:

$$CB = 0.447(0.684\,CB) + 653$$
$$= 0.306\,CB + 653$$
$$CB - 0.306\,CB = 653$$
$$CB = \frac{653}{0.694}$$
$$\underline{CB = 940 \text{ N } C}$$

from equation (2)

$$AB = 0.684 \times 940$$
$$\underline{AB = 643 \text{ N } T}$$

Vector Triangle

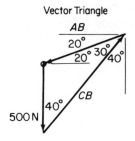

Figure 4-24

Method 2—Vector Triangle and Sine Law

Construct a vector triangle (Figure 4-24) from Figure 4-22 adding the vectors tip to tail until they close at the origin.

$$\frac{AB}{\sin 40°} = \frac{500}{\sin 30°}$$

$$AB = 500 \times \frac{0.643}{0.5}$$

$$AB = 643 \text{ N } T$$

$$\frac{CB}{\sin 110°} = \frac{500}{\sin 30°}$$

$$CB = 500 \frac{\sin 70°}{\sin 30°}$$

$$= 500 \times \frac{0.94}{0.5}$$

$$CB = 940 \text{ N } C$$

As you can see, the easier solution is usually the sine-law method. The important step in this method is the proper construction of the vector triangle.

Example 4-3

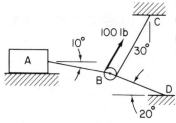

Weight A is pulled to the right by means of the ropes, a pulley, and a 100-lb force (Figure 4-25). Calculate the tension in ropes AB and BD.

Figure 4-25

Free–Body Diagram of B

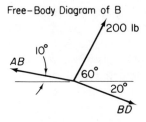

Figure 4-26

Since rope BC has a tension of 100 lb, there is a total applied force of 200 lb at B (Figure 4-26).

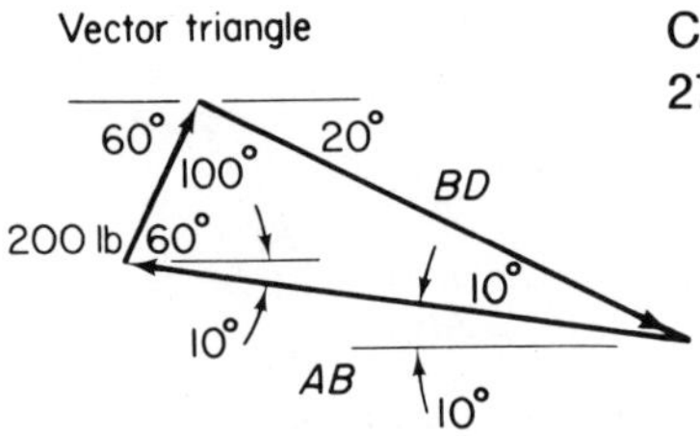

Vector triangle

Figure 4-27

Constructing the vector triangle (Figure 4-27) and applying the sine law, we have:

$$\frac{BD}{\sin 70°} = \frac{200}{\sin 10°}$$

$$BD = 200 \times \frac{0.94}{0.174}$$

$$BD = 1080 \text{ lb } T$$

$$\frac{AB}{\sin 100°} = \frac{200}{\sin 10°}$$

$$AB = 200 \times \frac{\sin 80°}{\sin 10°}$$

$$= 200 \times \frac{0.985}{0.174}$$

$$AB = 1130 \text{ lb } T$$

4-7 COPLANAR PARALLEL FORCE SYSTEMS

A horizontal beam with only vertical loading on it is an example of a force system in which all the forces are parallel and in the same plane—a *coplanar parallel system*. The beam is supported at two points. The free-body diagram replaces these supports with equivalent forces or reactions. One of the reaction forces is determined by taking moments about the other point of support.

Example 4-4

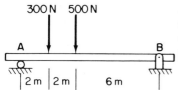

Figure 4-28

A beam has concentrated loads applied as shown in Figure 4-28. Calculate the reactions at A and B. (Neglect the weight of the beam.)

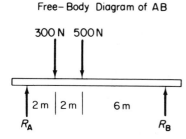

Free–Body Diagram of AB

Figure 4-29

The reaction at A, R_A, and the reaction at B, R_B, are indicated on the free-body diagram (Figure 4-29). Point B is pin connected and should also have a horizontal component, but, since there is no other horizontal force acting on the beam, there cannot be any horizontal component at B.

We now have a free-body diagram to which three equations may be applied;

$$\Sigma F_x = 0$$
$$\Sigma F_y = 0$$
$$\Sigma M = 0$$

If moments are taken about point A, R_A acts at a distance of zero from the center of moments and therefore drops out of the moment equation.

Stating the point about which moments are to be taken, we have:

$$\Sigma M_A = 0$$

or

counterclockwise moment − clockwise moment = 0
$$(R_B \times 10) - (300 \times 2) - (500 \times 4) = 0$$
$$10 R_B = 600 + 2000$$
$$\underline{R_B = 260 \text{ N} \uparrow}$$

Equating vertical forces, we obtain:

$$\Sigma F_y = 0$$

$$R_A + 260 - 300 - 500 = 0$$

$$R_A = 540 \text{ N} \uparrow$$

Since the value of R_A depends on R_B's having been calculated correctly, R_A can be checked by taking moments about point B.

$$\Sigma M_B = 0$$

$$-10 R_A + (500 \times 6) + (300 \times 8) = 0$$

$$10 R_A = 3000 + 2400$$

$$R_A = 540 \text{ N} \uparrow \text{ check}$$

Example 4-5

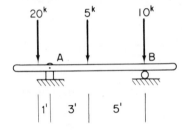

Solve for the reactions at points A and B (Figure 4-30).

Figure 4-30

Free–Body Diagram of AB

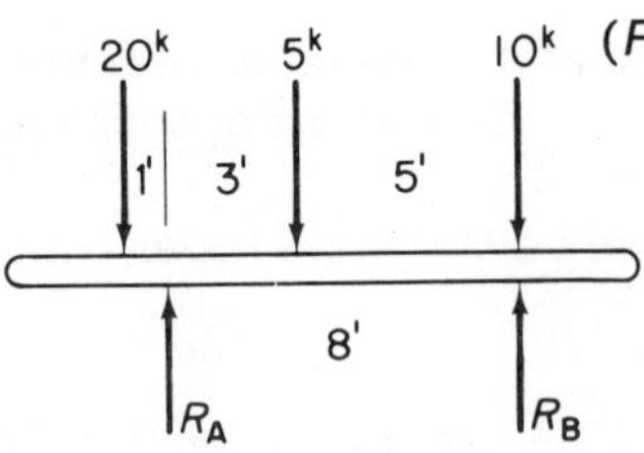

$$\Sigma M_A = 0 \qquad \text{(Figure 4-31)}$$

$$(R_B \times 8) + (20 \times 1) - (5 \times 3) - (10 \times 8) = 0$$

$$8 R_B + 20 - 15 - 80 = 0$$

$$8 R_B = 75$$

$$R_B = 9.4 \text{ kips} \uparrow$$

$$\Sigma F_y = 0$$

$$R_A + 9.4 - 20 - 5 - 10 = 0$$

$$R_A = 35 - 9.4$$

$$R_A = 25.6 \text{ kips} \uparrow$$

Figure 4-31

Instead of being subject to concentrated loads, a beam may have a *distributed load* applied along its length. The distributed load may be *uniform* or *nonuniform*. A distributed load can be visualized as the piling of concrete blocks along a beam's length either uniformly or to varying depths. The loading of a beam, whether it is uniform or nonuniform, is specified as weight per unit of length. In the English system, it may be lb/ft or kips/ft, and in the SI metric system it will be newtons per meter (N/m) or kN/m.

To determine the reactions where a beam is supported, the distributed load will be converted to a single force acting at the center of gravity of the distributed load. This assumption does not give the true bending effect on the beam, but it does allow us to solve for the reactions. The following examples show how to locate the center of gravity for various distributed loads.

Example 4-6

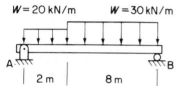

$W = 20$ kN/m $W = 30$ kN/m

2 m   8 m

Figure 4-32

Suppose that a beam is loaded at 20 kN/m over its first 2 m and at 30 kN/m over its remaining 8 m (Figure 4-32). Calculate R_A and R_B.

A concentrated force is assumed to be acting through the center of gravity of a distributed load.

The first distributed load we could convert would be 20 kN/m for 2 m, or $20 \times 2 = 40$ kN, acting at a distance of 1 m from A.

Similarly, 30 kN/m for 8 m gives a force of $30 \times 8 = 240$ kN acting 4 m from B.

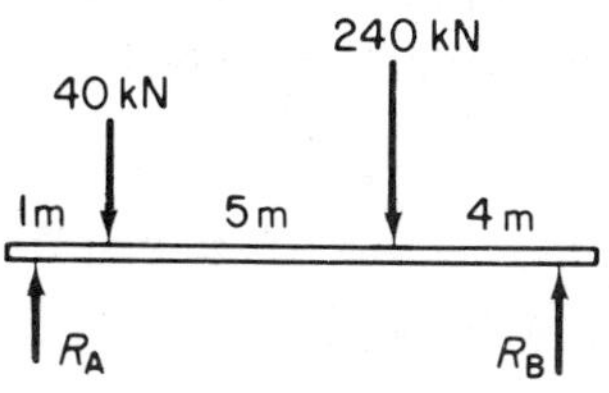

Free-Body Diagram of AB

240 kN

40 kN

Im   5m   4 m

R_A R_B

Figure 4-33

$\Sigma M_A = 0$ (Figure 4-33)

$10 R_B - (40 \times 1) - (240 \times 6) = 0$

$10 R_B = 40 + 1440$

$\underline{R_B = 148 \text{ kN} \uparrow}$

$\Sigma F_y = 0$

$R_A + 148 - 40 - 240 = 0$

$\underline{R_A = 132 \text{ kN} \uparrow}$

To solve for a more complex beam loading, we break it into simple loadings of either uniform or nonuniform loads.

In this way, the center of gravity and concentrated force of each are more easily found. A nonuniform load, triangular in shape, has a center of gravity located as shown in Figure 4-34. The concentrated force replacing this nonuniform load is essentially equal to the area of the triangle where we have units of $N/m \times m = N$.

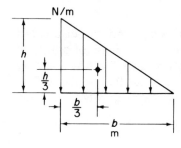

Figure 4-34

Example 4-7

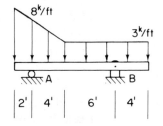

Figure 4-35

Solve for the reactions at A and B for the beam loaded as shown in Figure 4-35.

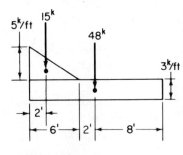

Figure 4-36

When converting the beam loading to equivalent concentrated forces, use as few forces as possible (Figure 4-36).

Each area in Figure 4-36 represents a force. The area of the triangle is $1/2 \times 6 \times 5$, or 15 kips. The rectangular area is $3 \times 16 = 48$ kips.

$$\Sigma M_A = 0 \qquad \text{(Figure 4-37)}$$

$$(R_B \times 10) - (48 \times 6) = 0$$

$$R_B = 28.8 \text{ kips} \uparrow$$

Free-Body Diagram of AB

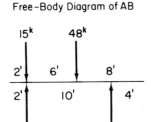

$\Sigma F_y = 0$

$R_A + 28.8 - 15 - 48 = 0$

$R_A = 63 - 28.8$

$\underline{R_A = 34.2 \text{ kips}\uparrow}$

Figure 4-37

Example 4-8

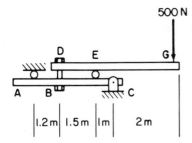

Figure 4-38

500 N The two beams in Figure 4-38 are fastened together by means of a bolt at D and B. Calculate the tension in the bolt and the reactions at A and C.

Free-Body Diagram of DG

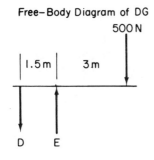

Figure 4-39

You have the choice of drawing free-body diagrams of DG or AC, or of both beams combined. In this case, we will start with beam DG since the known force is applied to this member.

$\Sigma M_E = 0$ \hfill (Figure 4-39)

$(D \times 1.5) - (500 \times 3) = 0$

$\underline{D = 1000 \text{ N}\downarrow}$

The bolt tension is thus 1 kN.

$\Sigma M_A = 0$ \hfill (Figure 4-40)

$(R_C \times 3.7) - (500 \times 5.7) = 0$

$\underline{R_C = 770 \text{ N}\uparrow}$

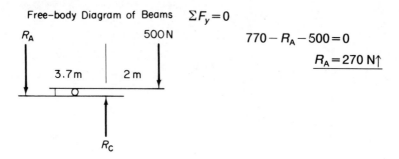

Free-body Diagram of Beams
R_A
500 N
3.7 m
2 m
R_C

$\Sigma F_y = 0$

$$770 - R_A - 500 = 0$$

$$\underline{R_A = 270 \text{ N}\uparrow}$$

Figure 4-40

4-8 COPLANAR NONCONCURRENT FORCE SYSTEMS

When dealing with *coplanar nonconcurrent force systems*, we are faced not only with vertical forces and moment equations but also with horizontal forces since the applied forces are no longer parallel.

Example 4-9

150 N
260 N
4
3
5
12
A
B
2 m
3m
1 m

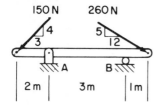

Figure 4-41

Solve for reactions R_A and R_B in Figure 4-41.

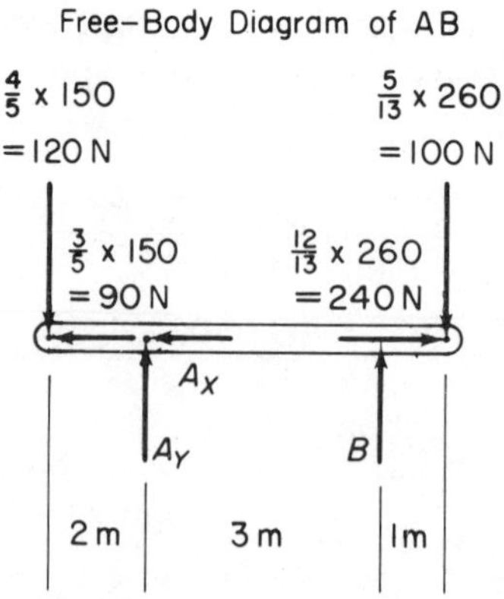

Free-Body Diagram of AB
$\frac{4}{5}$ x 150
= 120 N
$\frac{5}{13}$ x 260
= 100 N
$\frac{3}{5}$ x 150
= 90 N
$\frac{12}{13}$ x 260
= 240 N
A_X
A_Y
B
2 m
3 m
1 m

Figure 4-42

Resolve the 150- and 260-N forces into their horizontal and vertical components (Figure 4-42). Using $\Sigma F_y = 0$ would give an equation with two unknowns, A_y and B; instead, use either ΣM or $\Sigma F_x = 0$.

$\Sigma F_x = 0$

$$240 - 90 - A_x = 0$$

$$\underline{A_x = 150 \text{ N}\leftarrow}$$

Moments can be taken about A or B. Use point A since all horizontal forces and A_y have zero moments about A.

$$\Sigma M_A = 0$$

or

counterclockwise moment
$$-\text{clockwise moment} = 0$$

$$(120 \times 2) + (B \times 3) - (100 \times 4) = 0$$
$$3B = 400 - 240$$
$$\underline{B = 53.3 \text{ N}\uparrow}$$

$$\Sigma F_y = 0$$
$$A_y = 53.3 - 120 - 100 = 0$$
$$\underline{A_y = 167 \text{ N}\uparrow}$$

Example 4-10

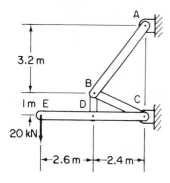

Find the reactions at A and C for the pin-connected structure shown in Figure 4-43.

Figure 4-43

Free–Body Diagram of Frame

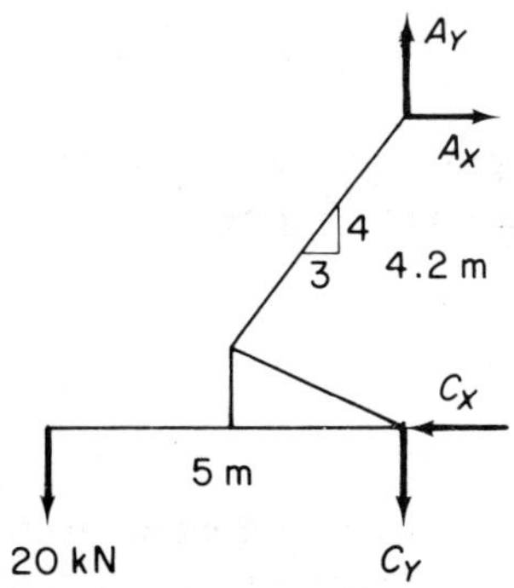

$$\Sigma M_C = 0 \qquad \text{(Figure 4-44)}$$
$$-(A_x \times 4.2) + (20 \times 5) = 0$$
$$\underline{A_x = 23.8 \text{ kN} \rightarrow}$$

Figure 4-44

Since AB is a two-force member, $A_x + A_y$ must be in the direction ⟋ ₄₃. Therefore:

$$A_y = \frac{4}{3} \times A_x$$

$$= \frac{4}{3} \times 23.8$$

$$\underline{A_y = 31.7 \text{ kN}\uparrow}$$

$$\Sigma F_x = 0$$

$$C_x = A_x$$

$$\underline{C_x = 23.8 \text{ kN}\leftarrow}$$

$$\Sigma F_y = 0$$

$$31.7 - C_y - 20 = 0$$

$$\underline{C_y = 11.7 \text{ kN}\downarrow}$$

Example 4-11

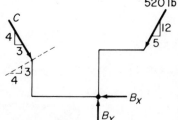

For the lever shown in Figure 4-45, solve for the pin reactions at B. Assume a smooth surface at C.

Figure 4-45

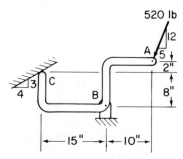

Free–Body Diagram of Lever

The free-body diagram may be constructed as in either Figure 4-46 or Figure 4-47.

Figure 4-46

Notice the slope of force C. We were given a slope of 3 to 4 for the smooth surface at C. The reaction force C is perpendicular to this surface and therefore has a slope of 4

Free-Body Diagram of Lever

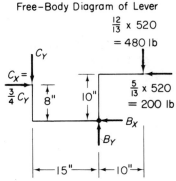

$\frac{12}{13}$ x 520

= 480 lb

$\frac{5}{13}$ x 520

= 200 lb

Figure 4-47

to 3. This is a rule that may be applied to all lines that are perpendicular to one another; that is, when the slope of one is known, merely reverse the numbers to find the slope of the other.

The free-body diagram in Figure 4-47 shows all forces as horizontal and vertical components.

The values of C_x and C_y must be found in order to obtain B_x and B_y; therefore, taking moments about point B (Figure 4-47), we get:

$$\Sigma M_B = 0$$

or

counterclockwise moment
$$-\text{clockwise moment} = 0$$

$$(C_y \times 15) + (200 \times 10) - \left(\frac{3}{4} C_y \times 8\right)$$

$$-(480 \times 10) = 0$$

$$15 C_y + 2000 - 6 C_y - 4800 = 0$$

$$9 C_y = 2800$$

$$C_y = 311 \text{ lb} \downarrow$$

$$C_x = \frac{3}{4} \times 311$$

$$= 233 \text{ lb} \rightarrow$$

$$\Sigma F_x = 0$$

$$233 - B_x - 200 = 0$$

$$B_x = 33 \text{ lb} \leftarrow$$

$$\Sigma F_y = 0$$

$$B_y - 311 - 480 = 0$$

$$B_y = 791 \text{ lb} \uparrow$$

Example 4-12

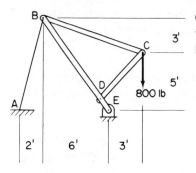

Calculate the horizontal and vertical reactions at point E of the frame shown in Figure 4-48. AB is a cable.

Figure 4-48

Note that it is preferable to show the cable tension components at A rather than at B since, when moments are taken about E, A_x passes through the pivot (Figure 4-49).

Free-Body Diagram of Frame

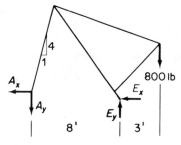

Figure 4-49

$$\Sigma M_E = 0$$

$$(A_y \times 8) - (800 \times 3) = 0$$

$$\underline{A_y = 300 \text{ lb} \downarrow}$$

$$A_x = \frac{1}{4} \times A_y$$

$$= \frac{1}{4} \times 300$$

$$\underline{A_x = 75 \text{ lb} \leftarrow}$$

$$\Sigma F_y = 0$$

$$E_y - 800 - 300 = 0$$

$$\underline{E_y = 1100 \text{ lb} \uparrow}$$

$$\Sigma F_x = 0$$

$$75 - E_x = 0$$
$$E_x = -75 \text{ lb} \leftarrow$$
$$\underline{E_x = +75 \text{ lb} \rightarrow}$$

Our calculation is still correct—only the direction of E_x must be changed.

PROBLEMS

4-1 Draw a free-body diagram of beam AC in Figure P4-1.

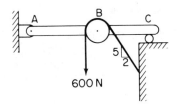

Figure P4-1

4-2 Draw a free-body diagram of pump handle AC in Figure P4-2.

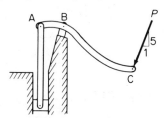

Figure P4-2

4-3 Draw a free-body diagram of member AC in Figure P4-3.

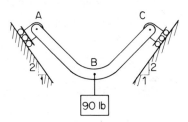

Figure P4-3

4-4   A jib crane supports an 80-lb load (Figure P4-4). Assume that all surfaces are smooth and draw a free-body diagram of the crane frame and label all forces.

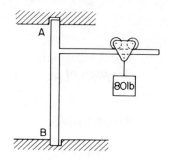

Figure P4-4

4-5   Draw free-body diagrams of members AB and CD in Figure P4-5.

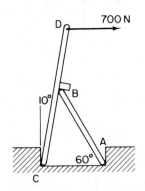

Figure P4-5

4-6   Draw a free-body diagram of member BD in Figure P4-6.

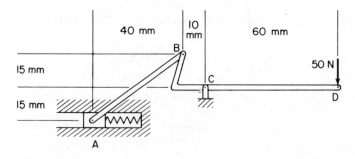

Figure P4-6

4-7 Draw a free-body diagram of member AC in Figure P4-7.

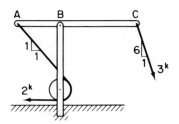

Figure P4-7

4-8 Draw free-body diagrams of members AC and DG in Figure P4-8.

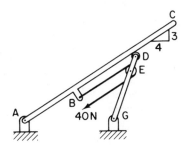

Figure P4-8

4-9 Rollers A and B each weigh 200 N (Figure P4-9). Assume smooth surfaces and draw a free-body diagram of roller A.

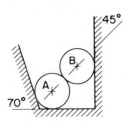

Figure P4-9

4-10 Draw a free-body diagram of the frame shown in Figure P4-10.

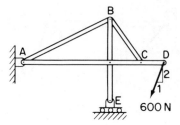

Figure P4-10

4-11 Two-force member CD has a compressive load of 2 kips when the frame in Figure P4-11 is loaded as shown. Draw a free-body diagram of member BE and label all horizontal and vertical components of forces acting on it. (Do not calculate actual values.)

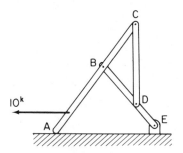

Figure P4-11

4-12 Draw a free-body diagram of member BD in Figure P4-12 and label the horizontal and vertical components of all forces acting on it. Show the components of *AC* as fractions of the total force in *AC*. (Do not calculate actual values).

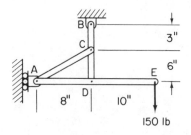

Figure P4-12

4-13 Draw a free-body diagram of member BF in Figure P4-13 and label all horizontal and vertical components. Show the components of *BC* as fractions of the total compressive force *BC*.

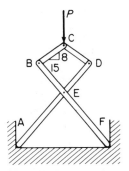

Figure P4-13

4-14 Draw a free-body diagram of member DE in Figure P4-14. Do not calculate values but label all forces acting on DE as horizontal and vertical components.

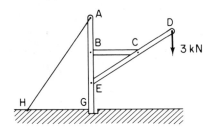

Figure P4-14

4-15 Draw a free-body diagram of member AE in Figure P4-15 and label all horizontal and vertical components. Show the components of *BC* and *DC* as fractions of the total load in each. (Do not calculate actual values.)

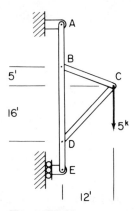

Figure P4-15

4-16 Draw a free-body diagram of member AD in Figure P4-16.

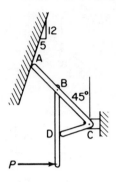

Figure P4-16

4-17 Draw free-body diagrams of members AC and BG in Figure P4-17.

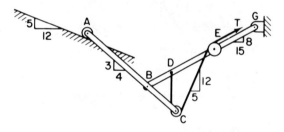

Figure P4-17

4-18 Determine the tension in both the cable and the spring in Figure P4-18.

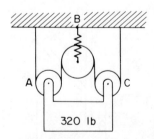

Figure P4-18

4-19 Determine the tension T in the cable for each pulley system in Figure P4-19.

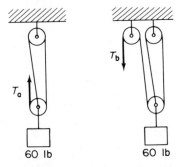

Figure P4-19a   **Figure P4-19b**

4-20 Determine the tension T in the cable for each pulley system in Figure P4-20.

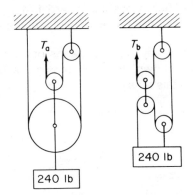

Figure P4-20a   **Figure P4-20b**

4-21 If A has a mass of 300 kg, determine the mass of B if the system in Figure P4-21 is to be in balance.

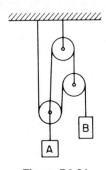

Figure P4-21

4-22 If the mass of B is 50 kg, determine the mass of A if the system in Figure P4-22 is to be in balance.

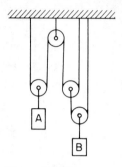

Figure P4-22

4-23 A belt is passed over two pulleys of equal diameter and then is tightened. Center-to-center distance between pulleys is 30 in. In checking the belt tension, the belt is pushed inward 1/4 in. by a force of 10 lb at a point midway between the pulleys. Calculate the belt tension.

4-24 The belt tension is adjusted by means of idler pulley A in Figure P4-24. At the position shown, the belt tension is 65 N. What is the moment about B due to this tension?

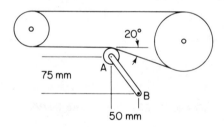

Figure P4-24

4-25 Determine the load in members AC and BC in Figure P4-25.

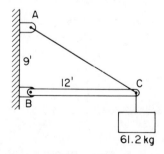

Figure P4-25

4-26 Determine the compressive loads in members AB and BC in Figure P4-26.

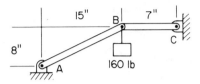

Figure P4-26

4-27 Determine the cable tension in each length of cable shown in Figure P4-27.

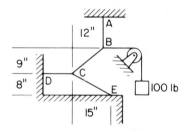

Figure P4-27

4-28 Determine the load in each pin-connected member in Figure P4-28.

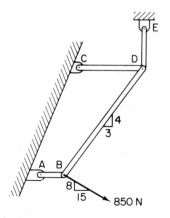

Figure P4-28

4-29 Determine the load in members AB and BC in Figure P4-29.

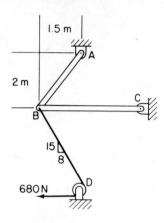

Figure P4-29

4-30 Determine the load in each length of cable in Figure P4-30.

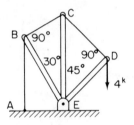

Figure P4-30

4-31 Determine the load in members AB and AC if force *P* is sufficient to hold the 3-kN force as shown in Figure P4-31.

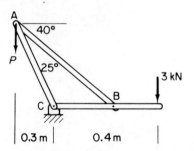

Figure P4-31

4-32 Assume smooth surfaces at points A and E in Figure P4-32. Determine the load in each member supporting the 3000-lb load.

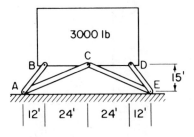

Figure P4-32

4-33 Each smooth roller in Figure P4-33 weighs 50 lb and has a diameter of 20 in. Calculate the reaction forces on the cylinder at points A and B.

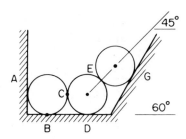

Figure P4-33

4-34 In Figure P4-34, beam AB is rotated from position A to position B by lengthening cable CB. Determine the cable tension *CB* for each position.

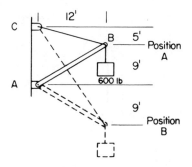

Figure P4-34

4-35 The cable and beam construction in Figure 4-35 supports a load of 2 kN. Determine the load in each beam and cable.

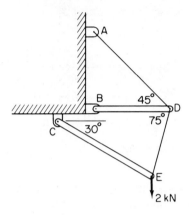

Figure P4-35

4-36 Determine the force P required to hold the 800-lb weight A by means of the system shown in Figure P4-36.

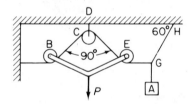

Figure P4-36

4-37 A 2-kN load is lifted by the cable system in Figure P4-37. Determine the tension T in the cable that passes over pulley A.

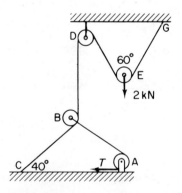

Figure P4-37

4-38–4-49 Determine the reactions at A and B for the beams loaded as shown in Figures P4-38 to P4-49. Beam weight may be neglected in all cases.

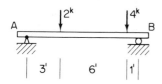

Figure P4-38

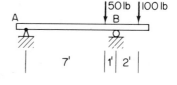

Figure P4-39

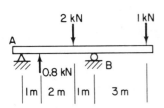

Figure P4-40

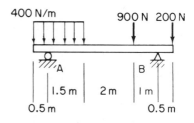

Figure P4-41

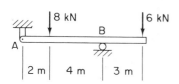

Figure P4-42

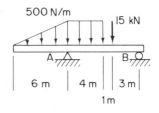

Figure P4-43

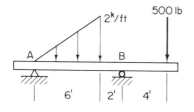

Figure P4-44

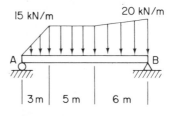

Figure P4-45

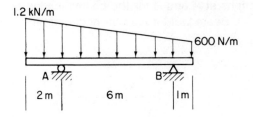

Figure P4-46

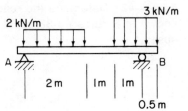

Figure P4-47

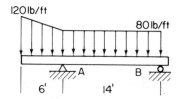

Figure P4-48

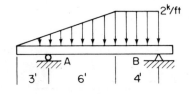

Figure P4-49

4-50 For the beam system in Figure P4-50, determine the cable tension *DE* and the reactions at A and B.

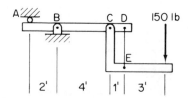

Figure P4-50

4-51 Two planks are bolted together by spacers in Figure P4-51. Determine the load in bolt C and the compressive load on spacer B.

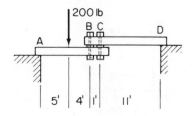

Figure P4-51

4-52 Determine the load in cable DC in Figure P4-52.

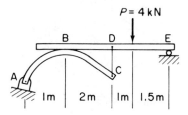

Figure P4-52

4-53 The tractor in Figure P4-53 weighs 5000 lb, the trailor weighs 4000 lb, and the load weighs 2000 lb. Determine the load on each set of wheels at A, B, and C.

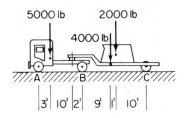

Figure P4-53

4-54 Find the reactions at A and B for the beam shown in Figure P4-54.

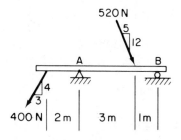

Figure P4-54

4-55 Find the reactions at A and B for the beam shown in Figure P4-55.

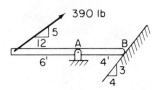

Figure P4-55

4-56 Find the cable tension and the reactions at A for the beam shown in Figure P4-56.

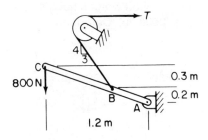

Figure P4-56

4-57 If the spring tension is 680 N in Figure P4-57, determine the reactions at B and C on member AC.

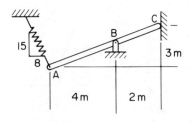

Figure P4-57

4-58 A storage rack is bolted to the wall at A and rests on the floor at B in Figure P4-58. Determine the reactions at A and B if the rack carries a total load of 900 lb, and the load has a center of gravity 2 ft from the wall.

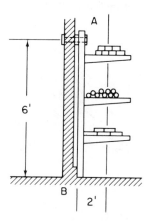

Figure P4-58

4-59 The frame shown in Figure P4-59 has a pulley at C that has a diameter of 2 ft. Determine the reaction on the frame at A and B. (The cable is parallel to AB.)

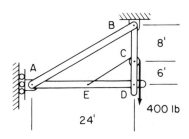

Figure P4-59

4-60 The light standard in Figure P4-60 is bolted to a concrete pedestal. The weight of the light, 10 lb, can be assumed to be 12 ft from the standard. The beam supporting the traffic lights weighs 150 lb (the weight can be assumed to be at the beam center), and each traffic light weighs 40 lb. Calculate the tension in bolts A and B if they are the only two installed.

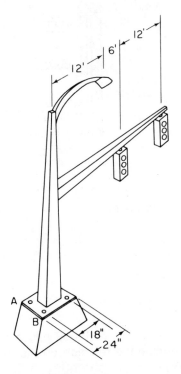

Figure P4-60

4-61 For the frame shown in Figure P4-61, determine the horizontal and vertical reactions at A and B.

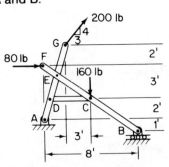

Figure P4-61

4-62 The tower shown in Figure P4-62 supports two loads of 2000 lb each due to wire weight. A side wind from the right causes the insulator cable AB to form a 10° angle with the vertical. Assume zero tension initially in the guy wires and calculate the tension in guy wire DE.

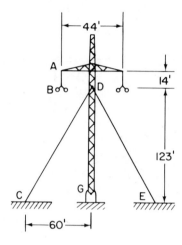

Figure P4-62

4-63 A gear box (Figure P4-63) receives an input torque of 12 N·m. Assume an output torque of the same amount. The gear box is mounted by means of bolts and spacers, which may exert a compressive or tensile force upon the gear box. Calculate the vertical force present at mounting points A, B, C, and D.

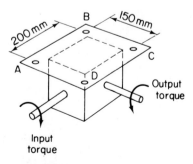

Figure P4-63

4-64 The frame shown in Figure P4-64 supports a mass A of 200 kg. Determine the reactions at C and G.

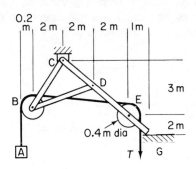

Figure P4-64

STRUCTURES AND MEMBERS

5-1 METHOD OF JOINTS

As mentioned previously, a *truss* is a structure composed of bars or members connected together to form one or more connected triangles. Each of the members is pinned at each end and, if carrying a load, is in either tension or compression. The direction of the member indicates the direction of the tensile or compressive force in the member acting on the joint. The ends of the members are pinned together to form a joint. A member in tension and pinned at a certain joint will be exerting a pull on the joint. A free-body diagram of this joint shows a vector acting away from the joint in the direction of the member. Each truss member is a two-force member.

The *method of joints* consists of a number of free-body diagrams of adjacent joints. The first joint selected must have only two unknown forces and one or more known forces. The unknown forces are determined by using $\Sigma F_x = 0$ and $\Sigma F_y = 0$. These newly found forces are used in the free-body diagram of an adjacent joint. The load in each truss member is found by taking consecutive free-body diagrams of joints throughout the complete truss.

Example 5-1

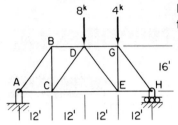

Figure 5-1

Determine the load in each member of the truss shown in Figure 5-1.

Free-Body Diagram of Truss

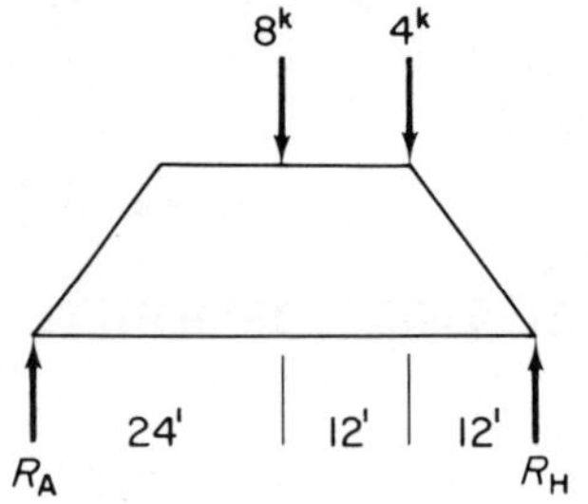

Figure 5-2

The first step of the solution is one with which you are already familiar: solve for the external reactions at points A and H (Figure 5-2).

Taking moments about point H and solving for R_A, we obtain:

$$\Sigma M_H = 0$$
$$(8 \times 24) + (4 \times 12) - (R_A \times 48) = 0$$
$$48 R_A = 192 + 48$$
$$R_A = \frac{240}{48}$$
$$\underline{R_A = 5 \text{ kips}\uparrow}$$

$$\Sigma F_y = 0$$
$$5 + R_H - 8 - 4 = 0$$
$$\underline{R_H = 7 \text{ kips}\uparrow}$$

In choosing the first joint of which you will draw a free-body diagram, notice that only four joints—A, D, G, and H—have known forces. Joint D has four unknowns, G has three unknowns, and joints A and H each has two unknowns. Thus, the first free-body diagram could be of joint A or H. Let us arbitrarily choose joint A.

Free–Body Diagram of A

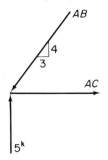

Figure 5-3

Considering member AB with its dimensions of 16 feet and 12 feet, we find the slope to be 4 to 3 (Figure 5-3).

Assume member AB to be in compression and member AC to be in tension. (Experience will make you more confident of these assumptions later.) The vectors are drawn so that member AB is pushing on the joint and member AC is pulling on it.

In considering vertical forces first, we find that the equation contains only one unknown, *AB*. The vertical component of *AB* is equal to 5 kips since there are no other vertical forces.

$$\Sigma F_y = 0$$

$$5 - \frac{4}{5}AB = 0$$

$$\underline{AB = 6.25 \text{ kips } C}$$

$$\Sigma F_x = 0$$

$$AC - \frac{3}{5}AB = 0$$

$$AC = \frac{3}{5} \times 6.25$$

$$\underline{AC = 3.75 \text{ kips } T}$$

The compression or tension of a member should be indicated following the value by *C* or *T* respectively.

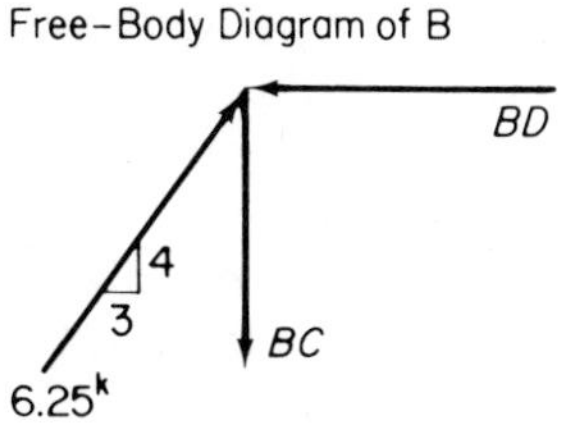

Free–Body Diagram of B

BD

4

3

BC

6.25ᵏ

Figure 5-4

We draw a free-body diagram of joint B next since it has only two unknowns (Figure 5-4).

AB was found to have a compression of 6.25 kips; therefore, 6.25 kips must be pushing on joint B even though vector direction is opposite to that in the free-body diagram of A.

$$\Sigma F_x = 0$$

$$\frac{3}{5} \times 6.25 - BD = 0$$

$$BD = 3.75 \text{ kips } C$$

$$\Sigma F_y = 0$$

$$\frac{4}{5} \times 6.25 - BC = 0$$

$$BC = 5 \text{ kips } T$$

Free-Body Diagram of C $\Sigma F_y = 0$ (Figure 5-5)

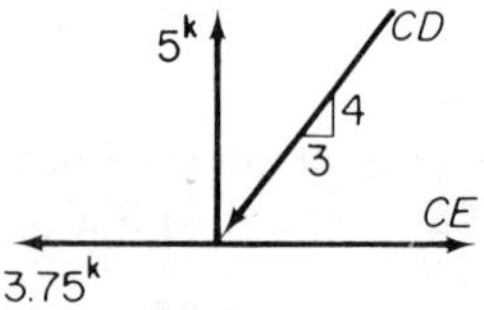

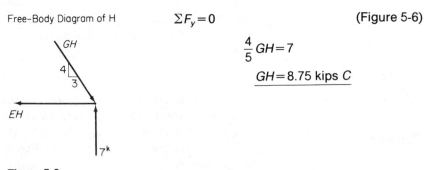

$$5 - \frac{4}{5} CD = 0$$

$$CD = 6.25 \text{ kips } C$$

$$\Sigma F_x = 0$$

Figure 5-5

$$CE - \frac{3}{5} CD - 3.75 = 0$$

$$CE = \left(\frac{3}{5} \times 6.25 \right) + 3.75$$

$$CE = 7.5 \text{ kips } T$$

At this point, approximately one-half of the truss member loads have been determined. Each calculated value depends on our having calculated the previous value correctly. If we now go to joint H and work back toward the center of the truss, the possibility of an error being perpetuated through the complete calculation is lessened. The problem solution is, in essence, broken into two halves, and we can conclude with a final check, using a free-body diagram of a joint near the center of the truss.

Free-Body Diagram of H $\Sigma F_y = 0$ (Figure 5-6)

$$\frac{4}{5} GH = 7$$

$$GH = 8.75 \text{ kips } C$$

Figure 5-6

$$\Sigma F_x = 0$$

$$\frac{3}{5}GH - EH = 0$$

$$\frac{3}{5} \times 8.75 - EH = 0$$

$$EH = 5.25 \text{ kips } T$$

Free-Body Diagram of G

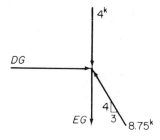

Figure 5-7

Do not forget to include the external 4-kip load in the free-body diagram of G (Figure 5-7).

$$\Sigma F_y = 0$$

$$\frac{4}{5} \times 8.75 - EG - 4 = 0$$

$$EG = 7 - 4$$

$$EG = 3 \text{ kips } T$$

$$\Sigma F_x = 0$$

$$DG - \frac{3}{5} \times 8.75 = 0$$

$$DG = 5.25 \text{ kips } C$$

Free-Body Diagram of E

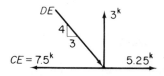

Figure 5-8

$$\Sigma F_y = 0 \qquad \text{(Figure 5-8)}$$

$$3 - \frac{4}{5}DE = 0$$

$$DE = 3.75 \text{ kips } C$$

$$\Sigma F_x = 0$$

$$\left(\frac{3}{5} \times 3.75\right) + (5.25) - 7.5 = 0$$

$$2.25 + 5.25 - 7.5 = 0$$

$$7.5 = 7.5 \qquad check$$

Since $CE = 7.5$ kips was calculated in the first half of the solution and now gives us a balance of horizontal forces at joint E, it would seem to indicate that all values calculated are correct.

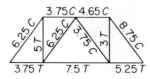

Figure 5-9

It is suggested that you conclude the solution with a sketch of the truss, labelling all member loads. Such a sketch is shown in Figure 5-9. This is useful not only as a final summary of all answers but, if you fill it in as you progress through the problem solution, also serves as a quick and easy way to find formerly calculated values when required for a new free-body diagram.

Example 5-2

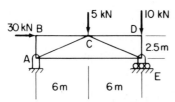

Figure 5-10

Use the method of joints to solve for the loads in members AB and CE of the truss shown in Figure 5-10.

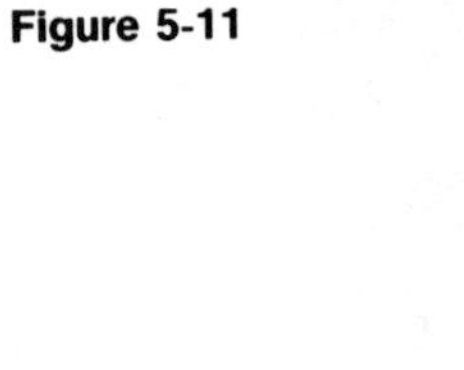

Free-Body Diagram of B

30 kN → ← BC

↓ AB

Figure 5-11

A free-body diagram of joint B (Figure 5-11) shows the load of member AB. Tension has been assumed, but, since there is no other vertical force present, the load in member AB is zero.

$$AB = 0$$

The load in member CE can be found from a free-body diagram of either joint C or E. The solution of joint C may be more difficult because of the joint's two members being sloped and unknown. Joint E will be easier to solve, but the load in member DE and the reaction at E will have to be found first.

Perhaps the best method of solution will not be readily apparent to you until you have solved several problems; so, until that time, you may simply have to draw many free-body diagrams and to use some trial-and-error methods. There is usually more than one method of solution, and experience will teach you the shortest one.

Solving for member DE, we have:

Free-Body Diagram of D

$$\Sigma F_y = 0 \qquad \text{(Figure 5-12)}$$

$$\underline{DE = 10 \text{ kN } C}$$

(Also note that the load in member CD equals zero.)

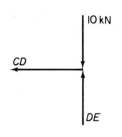

Figure 5-12

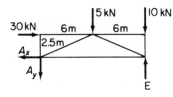

Free-Body Diagram of Truss

The reaction at E will be found from a free-body diagram of the complete truss (Figure 5-13).

$$\Sigma M_A = 0$$

$$(E \times 12) - (30 \times 2.5) - (5 \times 6)$$
$$- (10 \times 12) = 0$$
$$12E - 75 - 30 - 120 = 0$$
$$E = \frac{225}{12}$$
$$\underline{E = 18.75 \text{ kN} \uparrow}$$

Figure 5-13

Free-Body Diagram of E

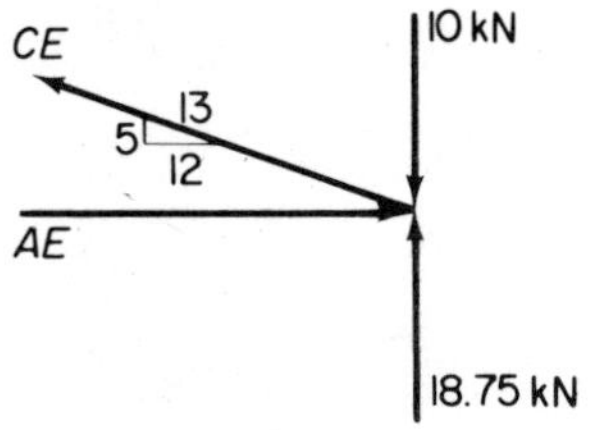

$$\Sigma F_y = 0$$

$$18.75 + \frac{5}{13} CE - 10 = 0$$

$$\frac{5}{13} CE = 10 - 18.75$$

$$CE = -8.75 \times \frac{13}{5}$$

$$\underline{CE = -22.8 \text{ kN } T}$$

Figure 5-14

Since this is a negative value, the direction assumed was incorrect; therefore, $CE = +22.8$ kN C.

Example 5-3

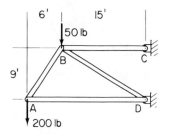

Figure 5-15

Solve for the load in each member of the pin-connected truss shown in Figure 5-15.

The vertical component of AB is $\dfrac{3}{3.61} \times AB$.

Free-Body Diagram of A

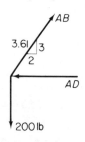

Figure 5-16

$\Sigma F_y = 0$ (Figure 5-16)

$$\frac{3}{3.61} AB - 200 = 0$$

$$\underline{AB = 240 \text{ lb } T}$$

$\Sigma F_x = 0$

$$\frac{2}{3.61} \times AB - AD = 0$$

$$\frac{2}{3.61} \times 240 - AD = 0$$

$$\underline{AD = 133 \text{ lb } C}$$

Free-Body Diagram of B

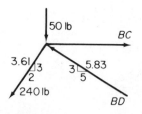

Figure 5-17

$\Sigma F_y = 0$ (Figure 5-17)

$$\frac{3}{5.83} BD - 50 - \frac{3}{3.61} \times 240 = 0$$

$$\frac{3}{5.83} BD - 50 - 200 = 0$$

$$BD = \frac{250 \times 5.83}{3}$$

$$\underline{BD = 486 \text{ lb } C}$$

$$\Sigma F_x = 0$$

$$BC - \left(\frac{5}{5.83} \times 486\right) - \left(\frac{2}{3.61} \times 240\right) = 0$$

$$BC - 417 - 133 = 0$$

$$\underline{BC = 550 \text{ lb } T}$$

Example 5-4

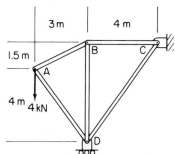

Figure 5-18

Solve for the load in each member of the system in Figure 5-18.

The solution generally begins with a free-body diagram of a joint on which a known force is acting. For this example, such a solution is not the easiest one, but we will first solve it in this fashion; an alternate solution will follow.

Free–Body Diagram of A

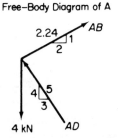

Figure 5-19

$$\Sigma F_y = 0 \qquad \qquad \text{(Figure 5-19)}$$

$$\left(\frac{1}{2.24} \times AB\right) + \left(\frac{4}{5} \times AD\right) - 4 = 0$$

$$0.447 AB + 0.8 AD = 4 \quad (1)$$

$$\Sigma F_x = 0$$

$$\frac{2}{2.24} AB - \frac{3}{5} AD = 0$$

$$1.49 AB = AD \qquad (2)$$

Substituting Equation (2) into Equation (1), we obtain:

$$0.447\,AB + 0.8\,(1.49\,AB) = 4$$
$$0.447\,AB + 1.19\,AB = 4$$
$$AB = \frac{4}{1.637}$$
$$\underline{AB = 2.44 \text{ kN } T}$$

Substitute $AB = 2.44$ into Equation (2):

$$AD = 1.49 \times 2.44$$
$$\underline{AD = 3.64 \text{ kN } C}$$

Free–Body Diagram of B

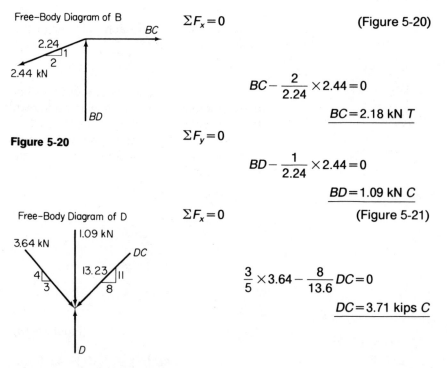

Figure 5-20

$\Sigma F_x = 0$ (Figure 5-20)

$$BC - \frac{2}{2.24} \times 2.44 = 0$$
$$\underline{BC = 2.18 \text{ kN } T}$$

$\Sigma F_y = 0$

$$BD - \frac{1}{2.24} \times 2.44 = 0$$
$$\underline{BD = 1.09 \text{ kN } C}$$

Free–Body Diagram of D

$\Sigma F_x = 0$ (Figure 5-21)

$$\frac{3}{5} \times 3.64 - \frac{8}{13.6}\,DC = 0$$
$$\underline{DC = 3.71 \text{ kips } C}$$

Figure 5-21

Simultaneous equations were used in this first solution; therefore, the alternate solution (which follows) may be preferred. The alternate solution begins with a free-body diagram of the frame to find the other external reactions, such as at C. The sequence of free-body diagrams of joints is then C, B, and A.

Example 5-4 (Alternate Solution)

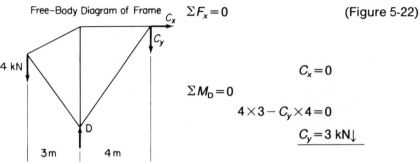

Free-Body Diagram of Frame

$\Sigma F_x = 0$ (Figure 5-22)

$C_x = 0$

$\Sigma M_D = 0$

$4 \times 3 - C_y \times 4 = 0$

$\underline{C_y = 3 \text{ kN} \downarrow}$

Figure 5-22

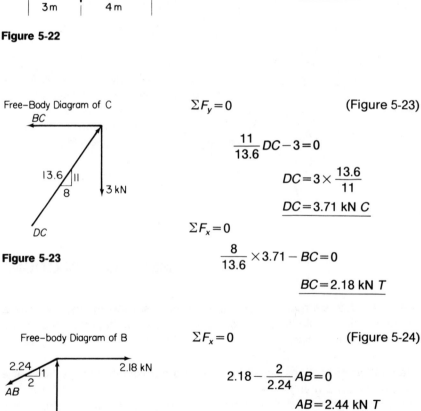

Free-Body Diagram of C

$\Sigma F_y = 0$ (Figure 5-23)

$\dfrac{11}{13.6} DC - 3 = 0$

$DC = 3 \times \dfrac{13.6}{11}$

$\underline{DC = 3.71 \text{ kN } C}$

$\Sigma F_x = 0$

$\dfrac{8}{13.6} \times 3.71 - BC = 0$

$\underline{BC = 2.18 \text{ kN } T}$

Figure 5-23

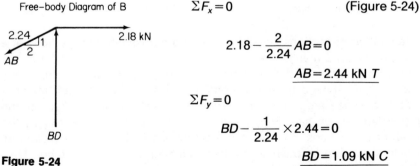

Free-body Diagram of B

$\Sigma F_x = 0$ (Figure 5-24)

$2.18 - \dfrac{2}{2.24} AB = 0$

$\underline{AB = 2.44 \text{ kN } T}$

$\Sigma F_y = 0$

$BD - \dfrac{1}{2.24} \times 2.44 = 0$

$\underline{BD = 1.09 \text{ kN } C}$

Figure 5-24

Free–Body Diagram of A

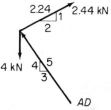

Figure 5-25

$\Sigma F_x = 0$ (Figure 5-25)

$$\frac{2}{2.24} \times 2.44 - \frac{3}{5} AD = 0$$

$$AD = 3.64 \text{ kN } C$$

Now check by taking:

$\Sigma F_y = 0$ (Figure 5-25)

$$\left(\frac{1}{2.24} \times 2.44\right) + \left(\frac{4}{5} \times 3.64\right) - 4 = 0$$

$$1.09 + 2.91 - 4 = 0$$

$$0 = 0 \qquad check$$

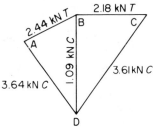

Figure 5-26

If, in using the analytical method (as in our second solution), we find that simultaneous equations can be avoided, a solution such as the alternate one is preferred. The load summary is shown in Figure 5-26.

5-2 METHOD OF SECTIONS

The *method of sections* is used when one is solving for the force in a member near the middle of a truss. The time-consuming method of joints is thereby unnecessary.

The method of sections consists of cutting a truss into two sections at a point at which a member force is required; one section is discarded. A free-body diagram of the remaining section is drawn. On this free-body diagram, we show a tensile or compressive force where each member was cut. These are equivalent forces that have the same effect as the discarded section had. Suppose that a truss member, cut in two by the method of sections, had been in compression. The free-body diagram would show a force pushing on the remaining half of the member. Only three members are usually cut at one time although a partial solution is possible when four or more members are cut.

Example 5-5

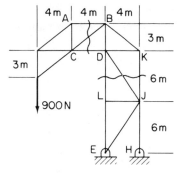

Figure 5-27

Solve for the load in members CB, AB, and JK of the truss shown in Figure 5-27.

A cutting plane is drawn through the truss, cutting members AB, CB, and CD. A free-body diagram of the left half of the truss is drawn since the external force of 900 N is acting on this section. The right half could be used, but then the reactions at E and H would have to be solved first.

Free–Body Diagram of Left Half

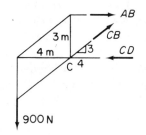

Figure 5-28

Member AB is assumed to be in tension, and vector *AB* is shown as a pull in the free-body diagram (Figure 5-28). Similarly, vector *CD* is pushing due to assumed compression. The nature of the load in member CB may not be so obvious, but it can be assumed to be in tension. The free-body diagram is now complete, and any one of the three equations of equilibrium can be applied.

$$\Sigma F_y = 0 \qquad \text{(Figure 5-28)}$$

$$\frac{3}{5} CB = 900$$

$$\underline{CB = 1500 \text{ N } T}$$

$$\Sigma M_C = 0$$

$$900 \times 4 - AB \times 3 = 0$$

$$\underline{AB = 1200 \text{ N } T}$$

Free-Body Diagram of Top Half

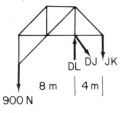

Figure 5-29

$$\Sigma M_D = 0 \qquad \text{(Figure 5-29)}$$

$$-(JK \times 4) + (900 \times 8) = 0$$

$$\underline{JK = 1800 \text{ N } T}$$

Example 5-6

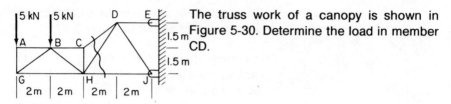

Figure 5-30

The truss work of a canopy is shown in Figure 5-30. Determine the load in member CD.

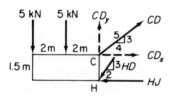

Free-Body Diagram of Left Half

Figure 5-31

The cutting plane is chosen such that it cuts only three members. Moments are taken about point H since the two other unknowns, HD and HJ, pass through this point (Figure 5-31). Rather than calculate the perpendicular distance between CD and point H, we can take horizontal and vertical components of *CD* at point C. Taking moments about point H, we find the only unknown is CD_x.

$$\Sigma M_H = 0 \qquad\qquad \text{(Figure 5-31)}$$

$$(5 \times 2) + (5 \times 4) - (CD_x \times 1.5) = 0$$

but

$$CD_x = \frac{4}{5} CD$$

$$1.5 \times \frac{4}{5} CD = 10 + 20$$

$$CD = 30 \times \frac{5}{6}$$

$$\underline{CD = 25 \text{ kN } T}$$

Example 5-7

For the truss loaded as shown in Figure 5-32, determine the load in member GE.

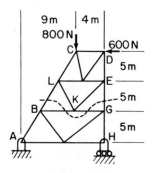

Figure 5-32

Free-Body Diagram of Top Half

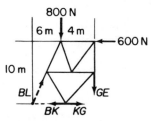

Figure 5-33

The cutting plane chosen (Figure 5-33) cuts four members. We will not be able to solve for all the members that are cut, but you will see that the load in member GE is quite readily found.

$$\Sigma M_B = 0 \qquad\qquad \text{(Figure 5-33)}$$
$$-(800 \times 6) + (600 \times 10) - (GE \times 10) = 0$$
$$\underline{GE = 120 NT}$$

Example 5-8

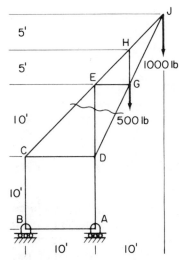

Figure 5-34

A crane has the truss framework shown in Figure 5-34. Determine the loads in members CE, DE, and DG.

A free-body diagram of the right half (Figure 5-35) is used to avoid solving for the reactions at A and B.

Moments must be used, and point J is chosen since the lines of action of two of the unknown forces pass through this point.

Free-Body Diagram of Right Half $\Sigma M_J = 0$ (Figure 5-35)

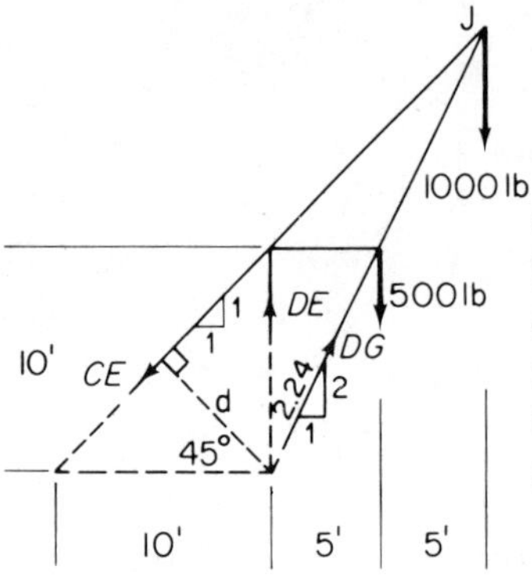

Figure 5-35

$$DE \times 10 = 500 \times 5$$
$$\underline{DE = 250 \text{ lb } C}$$

The use of horizontal or vertical forces would give simultaneous equations with two unknowns. Moments can be taken about point D, thus eliminating DE and DG from the moment equation and allowing us to find CE. The distance d, between CE and point D, will have to be calculated.

$$\cos 45° = \frac{d}{10}$$
$$d = 0.707 \times 10$$
$$d = 7.07 \text{ ft}$$

$\Sigma M_D = 0$
$$(CE \times 7.07) - (500 \times 5)$$
$$- (1000 \times 10) = 0$$
$$7.07\,CE = 2500 + 10,000$$
$$CE = \frac{12,500}{7.07}$$
$$\underline{CE = 1770 \text{ lb } T}$$

$\Sigma F_x = 0$
$$CE_x = DG_x$$
$$\cos 45° \times 1770 = \frac{1}{2.24} \times DG$$
$$1250 = \frac{DG}{2.24}$$
$$\underline{DG = 2800 \text{ lb } C}$$

5-3 METHOD OF MEMBERS

In the previous truss problems, the truss members were two-force members; there was a force acting at each end of the member; each member was in either tension or compression. The force of a member on a joint had the same direction as the slope of the member. For this reason, we were able to draw free-body diagrams of individual joints.

Free-body diagrams of joints cannot be used when the members have three or more forces acting on them. A three-force member may be subject to bending, and the force that it exerts on a joint no longer will have the same slope or direction as the member. Therefore, a free-body diagram is drawn of the member, not of the joint.

A *frame* consists of a number of members fastened together so that each member has two or more forces acting on it. Determining these forces consists of drawing free-body diagrams of individual members or of the complete frame.

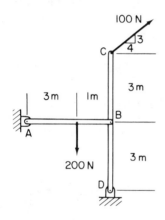

Figure 5-36

In Figure 5-36, each member has forces that tend to stretch and bend it. The free-body diagrams of each member are those shown in Figures 5-37 and 5-38. As before, any incorrectly assumed direction will give a negative answer of the correct magnitude.

Free-Body Diagram of AB

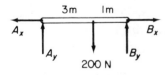

Figure 5-37

Note the direction of the vectors at B in both free-body diagrams. A free-body diagram of AB shows forces acting on AB. The force B_y on AB is the upward force of member CD on AB; that is, CD is holding AB up. Looking now at member CD, we notice that B_y is acting down because this is the direction in which AB acts on CD. The vector direction is opposite, but the force magnitude is the same. This is a rule that must always be followed when forces on pin-connected members are shown: whatever force direction is assumed at a point on a member, the opposite force direction must be shown at that point on the other connected member.

Free–Body Diagram of CD

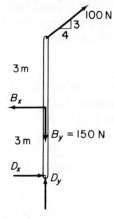

Figure 5-38

In the solution of the forces at B on member AB, one could begin with the free-body diagram of AB (Figure 5-37).

$$\Sigma M_A = 0$$

$$(B_y \times 4) - (200 \times 3) = 0$$

$$\underline{B_y = 150 \text{ N}\uparrow \text{ on } AB}$$

B_x cannot be obtained unless A_x is known. Use the free-body diagram of CD and take moments about D.

$$\Sigma M_D = 0 \qquad\qquad \text{(Figure 5-38)}$$

$$(B_x \times 3) - \left(\frac{4}{5} \times 100 \times 6\right) = 0$$

$$B_x = 160 \text{ N} \leftarrow \text{on CD}$$

or

$$\underline{B_x = 160 \text{ N} \rightarrow \text{on AB}}$$

Example 5-9

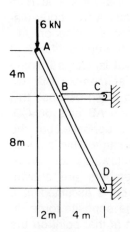

Figure 5-39

Determine the horizontal and vertical reactions at D (Figure 5-39).

Free-Body Diagram of AD

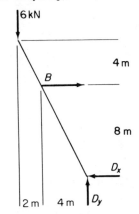

Figure 5-40

$\Sigma F_y = 0$ (Figure 5-40)

$\underline{D_y = 6 \text{ kN}\uparrow}$

$\Sigma M_B = 0$

$$(6 \times 4) + (6 \times 2) - (D_x \times 8) = 0$$
$$8D_x = 36$$
$$\underline{D_x = 4.5 \text{ kN}\leftarrow}$$

Example 5-10

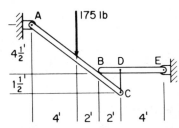

Figure 5-41

A load of 175 lb applied to the members shown in Figure 5-41 causes a tension of 400 lb in cable DC. Determine the reaction at B; assume a smooth surface. Determine the horizontal and vertical reactions at A and E.

Free-Body Diagram of BE

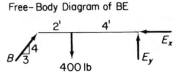

Figure 5-42

Since member AC has a slope of 3 to 4, force B perpendicular to it has a slope of 4 to 3 (Figure 5-42).

$\Sigma M_E = 0$

$$(400 \times 4) - \left(\frac{4}{5} B \times 6\right) = 0$$

$$B = \frac{5 \times 1600}{4 \times 6}$$

$$\underline{B = 333 \text{ lb}}$$

$$\Sigma F_x = 0$$

$$\frac{3}{5}B - E_x = 0$$

$$E_x = \frac{3}{5} \times 333$$

$$\underline{E_x = 200 \text{ lb} \leftarrow}$$

$$\Sigma F_y = 0$$

$$\left(\frac{4}{5} \times B\right) + E_y - 400 = 0$$

$$\left(\frac{4}{5} \times 333\right) + E_y = 400$$

$$E_y = 400 - 266$$

$$\underline{E_y = 134 \text{ lb} \uparrow}$$

Free–Body Diagram of Frame

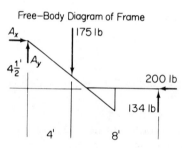

Figure 5-43

The reactions at A can be found by using a free-body diagram either of AC or of the complete frame—the latter is easier (Figure 5-43).

$$\Sigma F_x = 0$$

$$\underline{A_x = 200 \text{ lb} \rightarrow}$$

$$\Sigma F_y = 0$$

$$A_y + 134 - 175 = 0$$

$$\underline{A_y = 41 \text{ lb} \uparrow}$$

Example 5-11

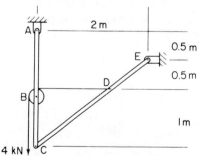

Figure 5-44

The frame shown in Figure 5-44 has a pulley 0.5 m in diameter at B. Determine the horizontal and vertical pin reactions at C.

Free–Body Diagram of AC

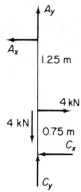

Figure 5-45

The cable tension of 4 kN is shown acting at the center of the pulley in Figure 5-45. Note the distances of 1.25 m and 0.75 m on AC.

There are too many unknowns in the horizontal and vertical directions; therefore, a moment equation must be used. Since C_x must be determined, moments are taken about A.

$$\Sigma M_A = 0$$

$$(4 \times 1.25) - (C_x \times 2) = 0$$

$$\underline{C_x = 2.5 \text{ kN} \leftarrow}$$

Free–Body Diagram of CE

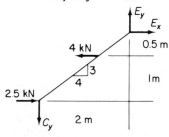

Figure 5-46

C_y cannot be determined from this diagram; so, a free-body diagram of CE (Figure 5-46) must be drawn. C_x and C_y are shown in a direction opposite to the one that they assume in the free-body diagram of AC.

$$\Sigma M_E = 0$$

$$(C_y \times 2) + (2.5 \times 1.5)$$

$$- (4 \times 0.5) = 0$$

$$2C_y = 2 - 3.75$$

$$C_y = -0.87 \text{ kN} \downarrow$$

$$\underline{C_y = 0.87 \text{ kN} \uparrow \text{ on CE}}$$

The magnitude of this answer is correct, but the direction of C_y was incorrectly assumed in both free-body diagrams.

Example 5-12

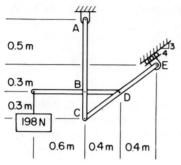

Figure 5-47

For the pin-connected frame shown in Figure 5-47, determine the horizontal and vertical reactions at B on AC.

Free-Body Diagram of BD

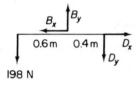

Figure 5-48

A free-body diagram of BD (Figure 5-48) contains the only known force and the horizontal and vertical reactions at B.

$$\Sigma M_D = 0$$

$$(198 \times 1) - (B_y \times 0.4) = 0$$

$$B_y = 495 \text{ N} \uparrow \text{on BD}$$

or

$$B_y = 495 \text{ N} \downarrow \text{on AC}$$

Free-Body Diagram of Frame

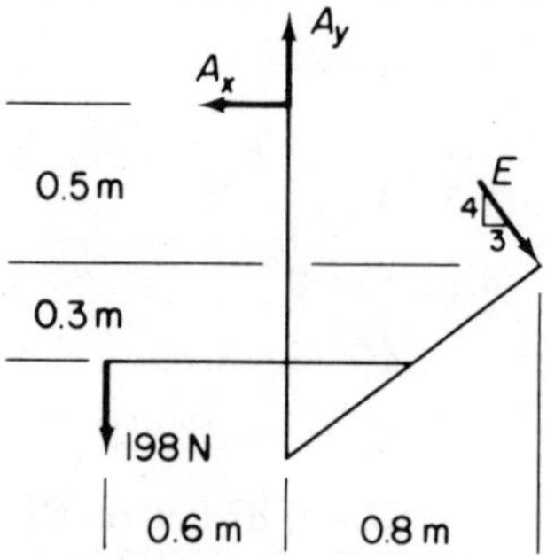

Figure 5-49

A free-body diagram of AC is indicated next, but it is found to have too many unknowns. Some of these unknowns, particularly those at A, can be found with a free-body diagram of the frame (Figure 5-49).

$$\Sigma M_A = 0$$

$$\left(\frac{4}{5}E \times 0.8\right) - (198 \times 0.6)$$

$$- \left(\frac{3}{5}E \times 0.5\right) = 0$$

$$0.64E - 0.3E = 119$$

$$E = 350 \text{ N} \quad ^{4} \underset{3}{\diagdown}$$

$$\Sigma F_x = 0$$

$$A_x = \frac{3}{5} \times E$$

$$A_x = \frac{3}{5} \times 350$$

$$\underline{A_x = 210 \text{ N} \leftarrow}$$

Free–Body Diagram of AC

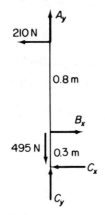

Figure 5-50

The value of $A_x = 210$ N is sufficient information to allow us to go to the free-body diagram of AC (Figure 5-50).

$$\Sigma M_C = 0$$

$$(210 \times 1.1) - (B_x \times 0.3) = 0$$

$$\underline{B_x = 770 \text{ N} \rightarrow \text{on AC}}$$

Example 5-13

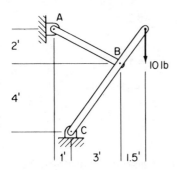

Figure 5-51

Solve for the horizontal and vertical pin reactions at A and C of the frame shown in Figure 5-51.

Free–Body Diagram of Frame

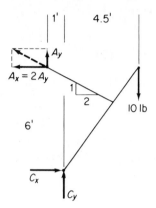

Figure 5-52

The key to the solution of this problem is to notice that member AB is a two-force member in tension. Referring to the free-body diagram of the frame (Figure 5-52), we note that the vector sum of A_x and A_y must have the same slope as member AB. Therefore

$$A_x = 2A_y$$

If moments are taken about point C, there is only one unknown, A_y.

$$\Sigma M_C = 0$$

$$(A_y \times 1) + (10 \times 4.5) - (2A_y \times 6) = 0$$
$$A_y + 45 - 12A_y = 0$$
$$A_y = \frac{45}{11}$$
$$\underline{A_y = 4.1 \text{ lb}\uparrow}$$

$$A_x = 2A_y$$
$$A_x = 2 \times 4.1$$
$$\underline{A_x = 8.2 \text{ lb}\leftarrow}$$

We can now calculate C_x and C_y by considering forces in the horizontal direction and forces in the vertical direction.

$$\Sigma F_x = 0$$
$$C_x = A_x$$
$$\underline{C_x = 8.2 \text{ lb}\rightarrow}$$

$$\Sigma F_y = 0$$
$$C_y + 4.1 = 10$$
$$\underline{C_y = 5.9 \text{ lb}\uparrow}$$

Example 5-14

Suppose that the frame in Figure 5-51 has an additional load of 20 lb applied as is

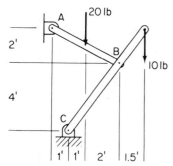

Figure 5-53

shown in Figure 5-53. Solve for the horizontal and vertical pin reactions at A and C in Figure 5-53.

Free-Body Diagram of Frame

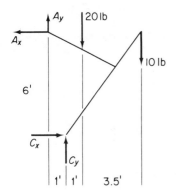

Figure 5-54

Both members of this frame are three-force members, and we do not know the relationship between the horizontal and vertical components at any point. In the free-body diagram of the frame (Figure 5-54) or that of AB (Figure 5-55), there are two unknowns regardless of which one of the three equilibrium equations is used. The method of solution requires selecting the same point in each diagram about which to take moments. The two moment equations will have the same two unknowns and will be solved simultaneously.

$$\Sigma M_C = 0 \qquad \text{(Figure 5-54)}$$

$$+(A_x \times 6) - (A_y \times 1) - (20 \times 1)$$

$$-(10 \times 4.5) = 0$$

$$6A_x - A_y = 65 \quad (1)$$

Free-Body Diagram of AB

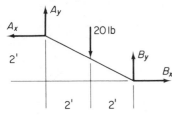

Figure 5-55

Considering the free-body diagram of AB, we note that moments about B will have A_x and A_y as unknowns.

$$\Sigma M_B = 0 \qquad \text{(Figure 5-55)}$$

$$-(A_y \times 4) + (A_x \times 2) + (20 \times 2) = 0$$

$$2A_x - 4A_y = -40 \quad (2)$$

Multiplying Equation (2) by -3 and adding Equation (1), we have:

$$-6A_x + 12A_y = 120 \qquad (2) \times -3$$
$$\underline{6A_x - A_y = 65 \qquad\qquad (1)}$$
$$11A_y = 185$$
$$\underline{A_y = 16.8 \text{ lb}\uparrow}$$

Substituting $A_y = 16.8$ into Equation (1), we get:

$$6A_x - 16.8 = 65$$
$$A_x = \frac{81.8}{6}$$
$$\underline{A_x = 13.6 \text{ lb}}$$

Referring now to the free-body diagram of the frame and solving for C_x and C_y, we have:

$$\Sigma F_x = 0 \qquad\qquad\qquad\qquad\qquad \text{(Figure 5-54)}$$
$$C_x = A_x$$
$$\underline{C_x = 13.6 \text{ lb}\rightarrow}$$

$$\Sigma F_y = 0$$
$$C_y + 16.8 - 20 - 10 = 0$$
$$\underline{C_y = 13.2 \text{ lb}\uparrow}$$

PROBLEMS

5-1–5-11 Use the method of joints to determine the force in each member of the trusses shown in Figures P5-1 to P5-11.

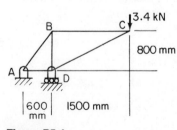

Figure P5-1

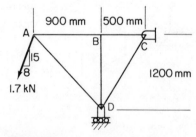

Figure P5-2

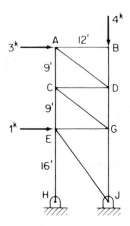

Figure P5-3

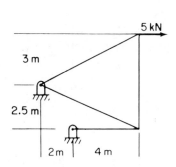

Figure P5-4

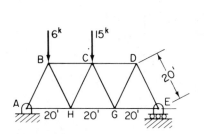

Figure P5-5

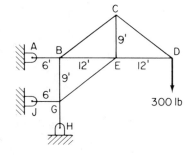

Figure P5-6

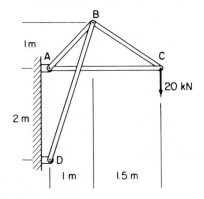

Figure P5-7

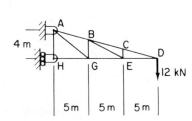

Figure P5-8

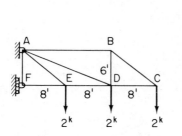

Figure P5-9

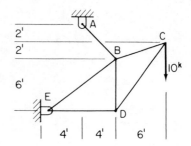

Figure P5-10

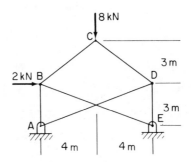

Figure P5-11

5-12 Determine the force in members BC and CG of the pin-connected truss shown in Figure P5-12.

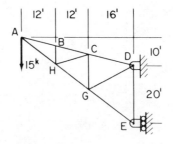

Figure P5-12

5-13 The scissor linkage shown in Figure P5-13 is controlled by cylinder CD and is used to compress material in a container below with a vertical force of 1000 lb. What cylinder force is required if the force of 1000 lb is being applied at the position shown?

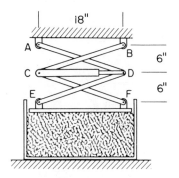

Figure P5-13

5-14 Use the method of joints to solve for the force in member BC of the truss shown in Figure P5-14.

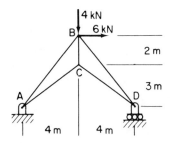

Figure P5-14

5-15 Determine the force in members BC, BG, and EG of the truss loaded as shown in Figure P5-15.

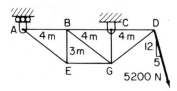

Figure P5-15

5-16 Determine the force in members CG and GJ of the truss shown in Figure P5-16.

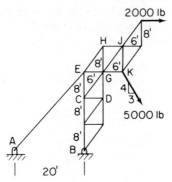

Figure P5-16

5-17 Determine the force in members BC, BE, BD, and DE of the truss shown in Figure P5-17.

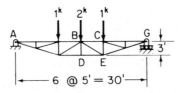

Figure P5-17

5-18 For the truss loaded as shown in Figure P5-18, determine the loads in members DE, CE, and CG.

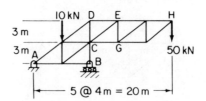

Figure P5-18

5-19 For the truss loaded as shown in Figure P5-19, determine the loads in members CD, EG, and EH.

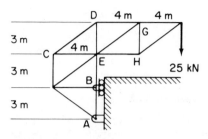

Figure P5-19

5-20 Determine the force in member EG of the K-truss shown in Figure P5-20.

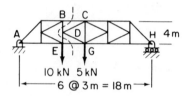

Figure P5-20

5-21 Use the method of sections to determine the loads in members CD and ED of the truss shown in Figure P5-21.

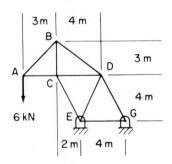

Figure P5-21

5-22 Determine the load in members BC, BH, and JH of the truss shown in Figure P5-22.

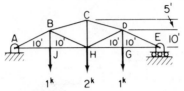

Figure P5-22

5-23 Determine the force in members DE, GE, and GH of the truss shown in Figure P5-23.

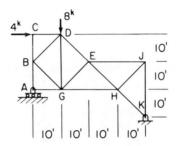

Figure P5-23

5-24 The truss framework of a billboard is subjected to the forces due to wind and the billboard weight as shown in Figure P5-24. Use the method of sections to determine the force in members CB, BE, and BG.

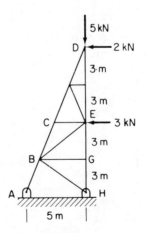

Figure P5-24

5-25 Use the method of sections to determine the load in members CD and CG of the saw-tooth truss shown in Figure P5-25.

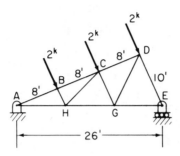

Figure P5-25

5-26 For the Parker truss shown in Figure P5-26, determine the force in members BC and BG.

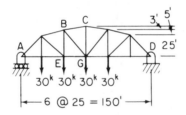

Figure P5-26

5-27 For the Fink truss shown in Figure P5-27, determine the load in members DE, JE, and KH. What is the load in members LM and MN?

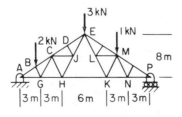

Figure P5-27

5-28 Use the method of sections to determine the load in members EC and ED of the truss shown in Figure P5-28.

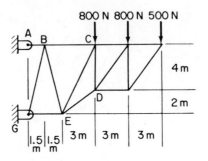

Figure P5-28

5-29 Determine the horizontal and vertical components of the pin reactions at B and D of the frame shown in Figure P5-29.

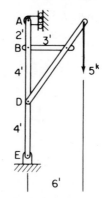

Figure P5-29

5-30 Determine the horizontal and vertical components of the pin reactions at B and C on member BE in Figure P5-30.

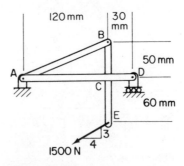

Figure P5-30

5-31 Determine the horizontal and vertical components of the pin reactions at B and C on member AC in Figure P5-31.

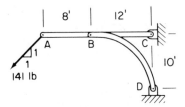

Figure P5-31

5-32 The frame in Figure P5-32 is held by cable HG when loaded as shown. Determine the horizontal and vertical components of the pin reactions at B and C.

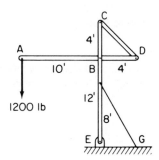

Figure P5-32

5-33 If a 5-kN load is applied as shown in Figure P5-33, determine the horizontal and vertical components of the pin reactions at B and C.

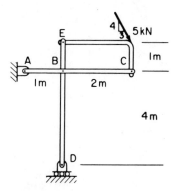

Figure P5-33

5-34 Determine the horizontal and vertical components of the pin reactions at B and D of the frame shown in Figure P5-34.

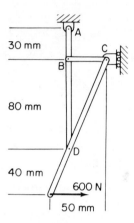

Figure P5-34

5-35 Determine the horizontal and vertical components of the pin reactions at B and C on member AD of the frame shown in Figure P5-35.

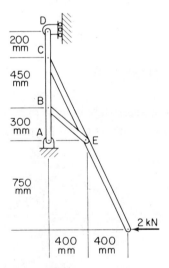

Figure P5-35

5-36 The platform supporting a 4.2-kip load in Figure P5-36 can be levelled by means of a cable and a winch at C. Neglect the weight of the platform and the drum diameter of the winch and calculate the tension in the cable and the pin reaction at B on the platform.

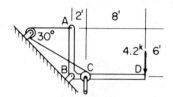

Figure P5-36

5-37 Bar AC of the frame shown in Figure P5-37 has a loading of 100 lb/ft. Determine the horizontal and vertical components of the pin reactions at A and D.

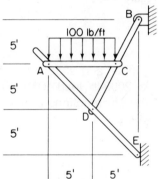

Figure P5-37

5-38 A 100-kg mass is lifted by means of the mechanism shown in Figure P5-38. Neglect the diameters of the pulleys and determine the horizontal and vertical components of the pin reaction at C on member ED.

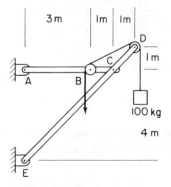

Figure P5-38

5-39   Determine the force *P* for the system shown in Figure P5-39 to be in static equilibrium. What are the horizontal and vertical components of the pin reactions at A and B?

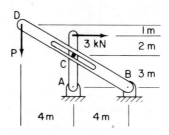

Figure P5-39

5-40   By means of a cylinder and slider at A, the structure shown in Figure P5-40 supports a 2000-lb load. Determine the horizontal and vertical components of the pin reactions at each point on member EB. (Neglect the pulley diameter at E.)

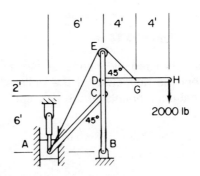

Figure P5-40

5-41   Determine the force required by the cylinder between D and H of the frame shown in Figure P5-41. (Neglect the roller diameter at J.)

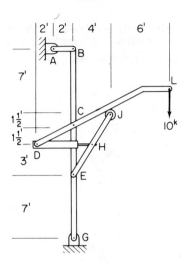

Figure P5-41

5-42 The toggle linkage shown in Figure P5-42 is used to clamp a work piece at G with a clamping force of 200 N. Determine (a) the pin reactions at D and C, and (b) the applied force *P*.

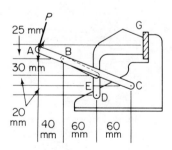

Figure P5-42

5-43 Pulley B is belt-driven by pulley A in Figure P5-43. By means of adjusting the turnbuckle DE, there is a belt tension of 400 N at pulley C. What is the tensile force in turnbuckle DE?

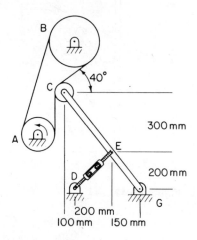

Figure P5-43

5-44 Material slid under the blade of the gap shear shown in Figure P5-44 requires 700 lb of vertical force to be sheared. Calculate the cylinder force required at D. Member AC pivots at point B.

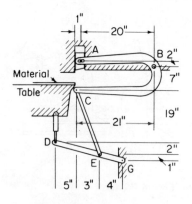

Figure P5-44

5-45 When two plates are spot welded together as in Figure P5-45, electrodes A and B squeeze the plates together with a force of 150 lb. What force must air cylinder EH apply in order to accomplish this?

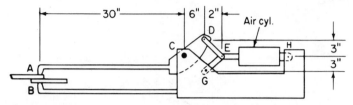

Figure P5-45

5-46 The side view of an evenly loaded clam-shell bucket is shown in Figure P5-46. The holding line and closing line have loads of 6 kN and 800 N, respectively. If the crosshead casting has a mass of 40 kg and the mass of arms AB and CD is neglected, determine the load in arms AB and CD.

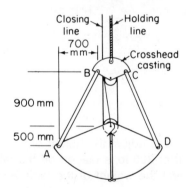

Figure P5-46

5-47 An automobile front-wheel assembly supports 3.5 kN. Determine the force compressing the spring and the components of the forces acting on the frame at points A and E (Figure P5-47).

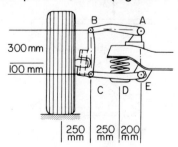

Figure P5-47

5-48 Find the force applied by each pair of cylinders of the loader when the bucket is positioned and loaded as shown in Figure P5-48. The lengths of CD and DF are 9 in. and 6 in., respectively. Member BC is parallel to AG.

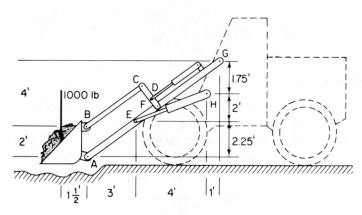

Figure P5-48

5-49 Cylinder BG is used to adjust the height of the scissor lift table in Figure P5-49. If a weight of 300 lb is placed on the table, what cylinder force is required? (Neglect the weight of the table and that of all other members.)

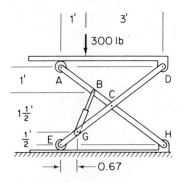

Figure P5-49

5-50 The gear puller shown in Figure P5-50 is used to pull gears and pulleys from a shaft by tightening the vertical screw. The screw pushes on the shaft with a vertical force of 800 N to remove the pulley shown. Assume smooth surfaces and determine the forces at B and E, and the forces in members AC and DG.

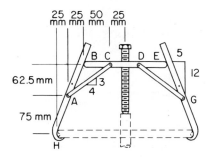

Figure P5-50

5-51 A tube may be bent around a roller as shown in Figure P5-51. Forces of 39 lb are applied at the positions indicated. Find the normal force at A and the pin reaction components at B on member BC. (Neglect the force due to friction at A.)

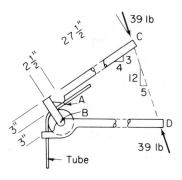

Figure P5-51

5-52 Members AC and BC each weighs 10 lb/ft and supports a 500-lb load as shown in Figure P5-52. Determine the horizontal and vertical components of the pin reactions at A and B.

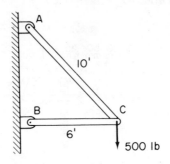

Figure P5-52

5-53 Determine the horizontal and vertical components of the pin reactions at A and B on the frame shown in Figure P5-53.

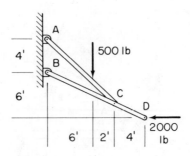

Figure P5-53

5-54 A cable is fastened at C and passes over a pulley at A. The cable tension is 6 kN. Neglect the pulley diameter and determine the pin reaction components at D and B (Figure P5-54).

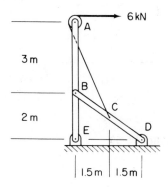

Figure P5-54

5-55 Determine the horizontal and vertical components of the pin reactions at C and D on the frame shown in Figure P5-55.

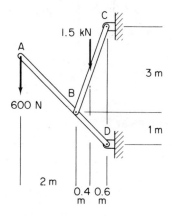

Figure P5-55

5-56 Determine the horizontal and vertical components of the pin reactions at D and C of the frame shown in Figure P5-56.

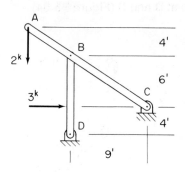

Figure P5-56

5-57 Determine the horizontal and vertical components of the reaction at A in Figure P5-57.

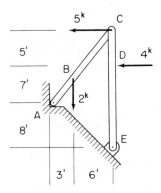

Figure P5-57

⑥

THREE-DIMENSIONAL
EQUILIBRIUM

6-1   RESULTANT OF PARALLEL FORCES

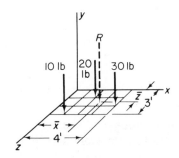

Figure 6-1

A system of parallel forces in three dimensions is shown in Figure 6-1. Gravity is the usual source of parallel forces, but there can be others, e.g., water pressure on a flat vertical wall. The location of the resultant of these parallel force systems must be found for applications such as the design of column foundation pads.

As in two-dimensional statics, a resultant is a single force that has the same effect as the system of forces that it replaces. In Figure 6-1, the magnitude of the resultant is equal to the algebraic sum of the vertical forces.

$$R = -10 - 20 - 30$$
$$= -60$$
$$R = 60 \text{ lb}\downarrow$$

One locates the resultant by determining the x and z coordinates and *designating them $\bar{x}$ and $\bar{z}$.* In three dimensional statics, *moments are always taken about an axis.* This fact may be more obvious in three-dimensional analysis than in two-dimensional analysis, where the axis appears to be a point. The forces shown in Figure 6-1 have moments about both the x- and z-axes, but not about the y-axis.

Using the sign convention of clockwise moments as negative and counterclockwise moments as positive, one can take moments about the x-axis. To visualize the moment direction, imagine yourself looking down the x-axis toward the origin. In the case in Figure 6-1, there are positive counterclockwise moments.

$$+ R\bar{z} = +(10 \times 3) + (20 \times 1) + (30 \times 2)$$
$$+60\bar{z} = +30 + 20 + 60$$
$$\bar{z} = \frac{110}{60}$$
$$\bar{z} = 1.83 \text{ ft}$$

Similarly, moments are taken about the z-axis.

$$-60\bar{x} = -(10 \times 1) - (20 \times 2) - (30 \times 4)$$
$$-60\bar{x} = -10 - 40 - 120$$
$$\bar{x} = \frac{-170}{-60}$$
$$\bar{x} = 2.83 \text{ ft}$$

Example 6-1

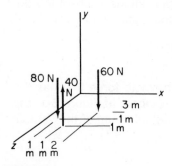

Determine the magnitude and location of the resultant of the force system shown in Figure 6-2.

Figure 6-2

The magnitude of R is equal to the algebraic sum in the y-direction.

$$R = -80 - 60 + 40$$
$$R = -100 \text{ N}$$

or

$$\underline{R = 100 \text{ N}\downarrow}$$

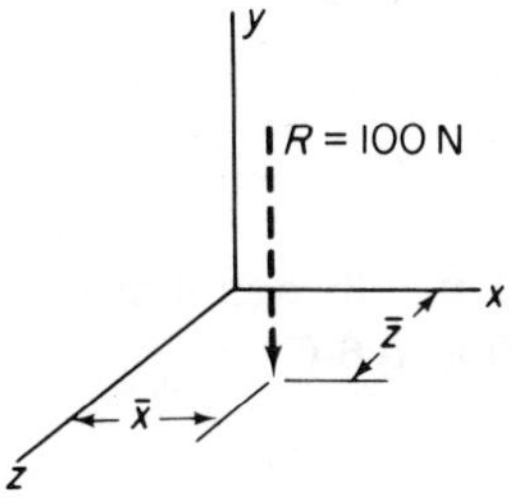

Take moments about the x-axis to determine the distance $\bar{z}$ from the x-axis to R (Figure 6-3).

$$+R \times \bar{z} = -(40 \times 5) + (80 \times 4) + (60 \times 3)$$
$$+100\bar{z} = -200 + 320 + 180$$
$$+100\bar{z} = +300$$
$$\underline{\bar{z} = 3 \text{ m}}$$

Similarly, taking moments about the z-axis, we get:

$$-R\bar{x} = (40 \times 2) - (80 \times 1) - (60 \times 4)$$
$$-100\bar{x} = 80 - 80 - 240$$
$$-100\bar{x} = -240$$
$$\underline{\bar{x} = 2.4 \text{ m}}$$

The magnitude and location of the resultant can then be shown (Figure 6-3).

Figure 6-3

Example 6-2

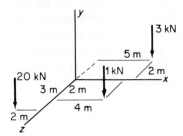

Figure 6-4

Determine the magnitude and location of the resultant of the force system shown in Figure 6-4.

$$R = -20 - 1 - 3$$
$$= -24$$
$$\underline{R = 24 \text{ kN}\downarrow}$$

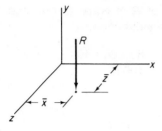

Figure 6-5

Assume that R is located as shown in Figure 6-5 and take moments about the x-axis.

$$+ R\bar{z} = +(20 \times 3) + (1 \times 2) - (3 \times 2)$$
$$24\bar{z} = 60 + 2 - 6$$
$$\underline{\bar{z} = 2.33 \text{ m}}$$

Taking moments about the z-axis, we get:

$$- R\bar{x} = +(20 \times 2) - (1 \times 4) - (3 \times 5)$$
$$-24\bar{x} = 40 - 4 - 15$$
$$\underline{\bar{x} = -0.875 \text{ m}}$$

The negative value of $\bar{x}$ indicates that R is located as shown in Figure 6-6.

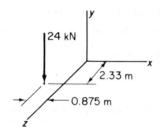

Figure 6-6

6-2   EQUILIBRIUM OF PARALLEL FORCES

The method followed here will allow the solution of only three unknowns. The direction and location of forces are known, so it becomes a matter of writing moment equations about the correct points or axes. The moment equations give simultaneous equations with two unknowns.

Example 6-3

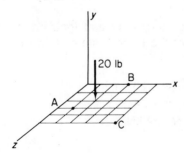

Figure 6-7

A horizontal plate is represented by the grid in Figure 6-7, where each square has sides 1 ft in length. The plate is supported at A, B, and C and has 20 lb applied as shown. Neglect the weight of the plate and determine the reactions at A, B, and C.

Although it may not be necessary in this case, moment equations may be more easily written if we draw a side view and a front view of the plate.

Side View — z-y Plane

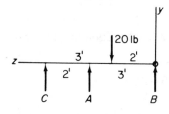

Figure 6-8

A side view (Figure 6-8) consists of looking down the x-axis toward the z-y-plane. This is often referred to as "projecting the forces into the z-y-plane."

Front View — x-y Plane

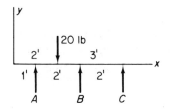

Figure 6-9

Similarly, a front view (Figure 6-9) consists of looking down the z-axis and viewing the forces in the x-y-plane.

Choose one of the three unknown forces—A, for example—and take moments about the point through which it passes in each diagram. Simultaneous equations with the unknowns B and C are then solved.

For the side view (Figure 6-8):

$$\Sigma M_A = 0$$

$$3B - (20 \times 1) + (C \times 2) = 0$$
$$3B - 2C = 20 \qquad (1)$$

For the front view (Figure 6-9):

$$\Sigma M_A = 0$$

$$(B \times 2) + (C \times 4) - (20 \times 1) = 0$$
$$2B + 4C = 20 \qquad (2)$$

Multiplying Equation (1) by 2 and adding

Equation (2), we get:

$$6B - 4C = 40 \qquad (1) \times 2$$
$$\underline{2B + 4C = 20} \qquad (2)$$
$$8B + 0 = 60$$
$$\underline{B = 7.5 \text{ lb} \uparrow}$$

Substituting $B = 7.5$ into Equation (1), we get:

$$3 \times 7.5 - 2C = 20$$
$$-2C = -2.5$$
$$\underline{C = 1.25 \text{ lb} \uparrow}$$

The summation of vertical forces will give the value of the third unknown, A.

$$\Sigma F_y = 0$$

$$A + B + C - 20 = 0$$
$$A + 7.5 + 1.25 - 20 = 0$$
$$\underline{A = 11.25 \text{ lb} \uparrow}$$

6-3   COMPONENTS AND RESULTANTS OF FORCES IN SPACE

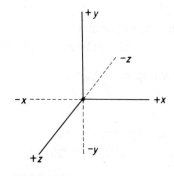

Figure 6-10

When describing the direction of a force in three dimensions, one must use the proper sign convention. The sign convention for each of the x-, y-, and z-axes is shown in Figure 6-10.

Let us find the resultant of three forces: $+4$ N in the x-direction; $+3$ N in the y-direction; and $+2$ N in the z-direction.

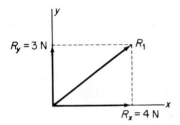

Figure 6-11

Add the *x* and *y* components to obtain a resultant, R_1. *This is precisely what we have been doing in all the coplanar force systems up to now.* From Figure 6-11, we have:

$$R_1 = \sqrt{(3)^2 + (4)^2}$$
$$R_1 = 5 \text{ N}$$

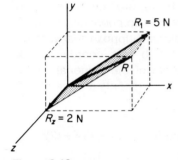

Figure 6-12

R_1 can now be added to the *z* component, $R_z = 2$ N, to obtain the final resultant, *R*. Note the shaded plane formed by these two forces in Figure 6-12.

$$R = \sqrt{(5)^2 + (2)^2}$$
$$R = 5.38 \text{ N}$$

R is the diagonal of a rectangular box formed by the three components.

Rather than adding the components in two steps, one could have solved for *R* in one step.

$$R = \sqrt{(R_x)^2 + (R_y)^2 + (R_z)^2}$$
$$= \sqrt{(4)^2 + (3)^2 + (2)^2}$$
$$R = 5.38 \text{ N}$$

The direction of this resultant is indicated by showing the *x, y, z* coordinates after the answer. The coordinates in this case are (4,3,2), and the final answer appears as:

$$\underline{R = 5.38 \text{ N, coordinates } (4,3,2)}$$

Example 6-4

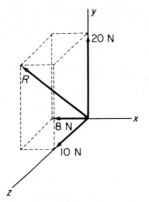

Figure 6-13

Determine the resultant of the forces shown in Figure 6-13.

$$R = \sqrt{(8)^2 + (20)^2 + (10)^2}$$
$$= \sqrt{564}$$
$$R = 23.7 \text{ N} (-4, 10, 5)$$

The coordinates $(-8, 20, 10)$ can be reduced to $(-4, 10, 5)$.

We now come to the problem of resolving a force in space into components in the x-, y-, and z-directions. Suppose that a force in space of 100 lb has coordinates of $(8, 4, 2)$. The length of the diagonal of the box (Figure 6-14) represents a force of 100 lb.

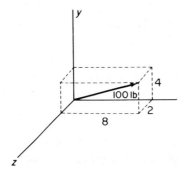

Figure 6-14

$$\text{diagonal length} = \sqrt{(8)^2 + (4)^2 + (2)^2}$$
$$= \sqrt{64 + 16 + 4}$$
$$= \sqrt{84}$$
$$= 9.16$$

By proportion, if a diagonal length of 9.16 represents 100 lb, the x dimension of the box (8) represents $8/9.16 \times 100 = 87.4$ lb. Therefore:

$$R_x = 87.4 \text{ lb} \rightarrow$$

Similarly:

$$R_y = \frac{4}{9.16} \times 100$$
$$R_y = 43.7 \text{ lb} \uparrow$$
$$R_z = \frac{2}{9.16} \times 100$$
$$R_z = 21.8 \text{ lb} \swarrow$$

This is the same principle that we applied in two-dimensional coplanar force systems.

A force R with a 3 to 4 slope has a horizontal component of:

$$R_x = \frac{4}{5}R$$

We are now applying the same principle to the third component, R_z. The following equation is a more concise method of showing the relationship among the force components.

$$\frac{R_x}{x} = \frac{R_y}{y} = \frac{R_z}{z} = \frac{R}{\sqrt{x^2 + y^2 + z^2}} \qquad (6\text{-}1)$$

or

$$\frac{R_x}{8} = \frac{R_y}{4} = \frac{R_z}{2} = \frac{100}{9.16}$$

$$R_x = 87.4 \text{ lb} \rightarrow, \ R_y = 43.7 \text{ lb} \uparrow,$$

and $R_z = 21.8$ lb ↙

Example 6-5

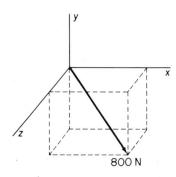

Figure 6-15

Determine the x, y, and z components of an 800-N force in space that has coordinates $(5, -12, 16)$ (Figure 6-15).

Diagonal length $= \sqrt{(5)^2 + (12)^2 + (16)^2}$

$$= \sqrt{25 + 144 + 256}$$

$$= \sqrt{425}$$

$$= 20.6$$

$$R_x = \frac{5}{20.6} \times 800$$

$$\underline{R_x = 194 \text{ N} \rightarrow}$$

$$R_y = \frac{12}{20.6} \times 800$$

$$\underline{R_y = 466 \text{ N} \downarrow}$$

$$R_z = \frac{16}{20.6} \times 800$$

$$\underline{R_z = 621 \text{ N} \swarrow}$$

6-4   EQUILIBRIUM IN THREE DIMENSIONS

Forces in space are noncoplanar, but may be

1. parallel,
2. concurrent, or
3. nonconcurrent.

(We have just finished considering parallel forces in space in Sections 6-1 and 6-2.)

The compressive loads in the legs of a camera tripod are an example of *concurrent forces.* These forces are concurrent because they all intersect the camera at a common point.

When the forces in three dimensions do not intersect at a common point, they form a noncoplanar, nonconcurrent force system. These are considered in more advanced designs when various member loads are required and where the reactions at supports can be more involved since there may be couples present. Only the simpler reactions with stated assumptions will be considered here. The regular equilibrium equations for forces in three dimensions will be used.

What are the equilibrium equations for concurrent forces in three dimensions? In coplanar force systems, we had three equations, and moments were taken about some point on the free-body diagram.

$$\Sigma F_x = 0$$
$$\Sigma F_y = 0$$
$$\Sigma M = 0$$

For forces in three dimensions, we have:

$$\Sigma F_x = 0 \quad \Sigma M_x = 0$$
$$\Sigma F_y = 0 \quad \Sigma M_y = 0$$
$$\Sigma F_z = 0 \quad \Sigma M_z = 0$$

Moments are taken about the *x*-, *y*-, and *z*-axes or about any other axis that may be convenient in the problem solution. There is a difference in moment equations; in a two-dimensional coplanar force system, moments can be considered as being taken about a point; in a three-dimensional noncoplanar force system, moments are taken about an axis.

There is more than one method of problem solution. The shortest one for some of the simpler problems consists of drawing a three-dimensional free-body diagram—somewhat like an isometric view—and applying the equilibrium equations. All forces are resolved into their three components and, if possible, are expressed as fractions of the total force.

If you become confused with all the forces shown on one diagram, the following three steps may simplify the problem solution:

Step 1   Project all forces into two or more planes, i.e., take front, side, or top views. Any force that has a line of action in the direction that you are viewing is not shown in that view. Another way of expressing this rule is: a force produces no reaction in a plane that is perpendicular to its line of action.

Step 2   If possible, show all these projected forces or components at any point as fractions of the total force at that point.

Step 3   Treat each view as a coplanar force system and apply the three equilibrium equations as before. Moment equations are also often used.

Example 6-6

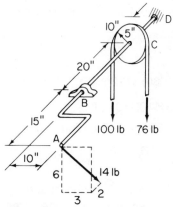

The crank in Figure 6-16 has a smooth bearing at B and a ball and socket at D. Calculate all reaction components at B and D.

Figure 6-16

Free-Body Diagram of Crank

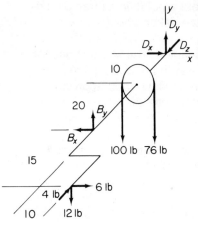

Figure 6-17

The three-dimensional free-body diagram in Figure 6-17 is used here. The rectangular components at A can be found if the diagonal length of the space coordinates is known.

$$\text{diagonal length} = \sqrt{(6)^2 + (3)^2 + (2)^2}$$

$$= 7$$

$$A_y = \frac{6}{7} \times 14 = 12 \text{ lb}$$

$$A_x = \frac{3}{7} \times 14 = 6 \text{ lb}$$

$$A_z = \frac{2}{7} \times 14 = 4 \text{ lb}$$

As stated before, when using a free-body diagram in three dimensions, one must take moments about an axis, not about a point. Moments about the y-axis would give us B_x. (All vertical forces have zero moment about the y-axis.)

$$\Sigma M_y = 0$$

$$-(B_x \times 30) + (6 \times 45) + (4 \times 10) = 0$$

$$30 B_x = 270 + 40$$

$$\underline{B_x = 10.3 \text{ lb} \leftarrow}$$

Take moments about the x-axis to obtain B_y.

$$\Sigma M_x = 0$$

$$-(B_y \times 30) + (12 \times 45)$$

$$+ (176 \times 10) = 0$$

$$30 B_y = 540 + 1760$$

$$\underline{B_y = 76.6 \text{ lb} \uparrow}$$

The components at D can be found by summation of forces in each of the three directions—x, y, and z.

$$\Sigma F_x = 0$$

$$D_x + 6 - B_x = 0$$
$$D_x = 10.3 - 6$$
$$\underline{D_x = 4.3 \text{ lb} \rightarrow}$$

$$\Sigma F_y = 0$$

$$D_y + B_y - 176 - 12 = 0$$
$$D_y = 176 + 12 - 76.6$$
$$\underline{D_y = 111.4 \text{ lb} \uparrow}$$

$$\Sigma F_z = 0$$

$$\underline{D_z = 4 \text{ lb} \swarrow}$$

Example 6-7

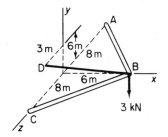

Figure 6-18

Determine the tensile load in cable DB and the compressive loads in members AB and BC (Figure 6-18). Neglect the weight of the members.

We will solve this problem by the view method of projecting the forces on a plane.

Find the length of diagonal DB.

$$DB = \sqrt{(6)^2 + (6)^2 + (3)^2}$$
$$= 9 \text{ m}$$

A top view and a side view can be drawn. The top view is obtained by looking down the y-axis and projecting all the forces onto the x-z-plane (Figure 6-19). The side view consists of looking along the z-axis toward the origin and projecting all the forces onto the x-y-plane (Figure 6-20).

Note that in the top view (Figure 6-19) the 3-kN force is not shown. In the side view

Top View or x-z Plane

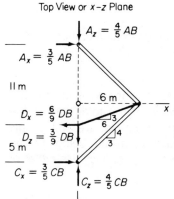

Figure 6-19

Side View or x–y Plane

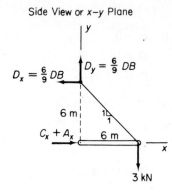

Figure 6-20

(Figure 6-20), there are no vertical components at points A and C since the weight of the members was neglected. Since the side view shows the given force of 3 kN, we can start here by taking moments about C.

$$\Sigma M_C = 0$$

$$\frac{6}{9} \times DB \times 6 - (3 \times 6) = 0$$

$$\underline{DB = 4.5 \text{ kN } T}$$

Use $DB = 4.5$ kN in the top view (Figure 6-19) and take moments about point A and then about point C.

$$\Sigma M_A = 0$$

$$-\left(\frac{6}{9} \times DB \times 11\right) + \left(\frac{3}{5} CB \times 16\right) = 0$$

$$\frac{6}{9} \times 4.5 \times 11 = \frac{3}{5} \times 16 \times CB$$

$$\underline{CB = 3.44 \text{ kN } C}$$

$$\Sigma M_C = 0$$

$$\left(\frac{6}{9} \times 4.5 \times 5\right) - \left(\frac{3}{5} \times AB \times 16\right) = 0$$

$$\underline{AB = 1.56 \text{ kN } C}$$

Example 6-8

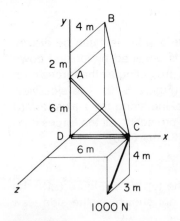

Figure 6-21

Determine the loads in all members shown in Figure 6-21.

In Example 6-7, we showed each component as a fraction of the total load in the member. In this example, we will simply label each component as being in the x-, y-, or z-direction. You may use either of the two methods.

Top View or *x-z* Plane

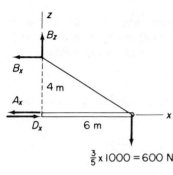

Figure 6-22

The top view (Figure 6-22) has the fewest unknowns; taking moments about A will give us B_x:

$$\Sigma M_A = 0$$

$$(B_x \times 4) - (600 \times 6) = 0$$

$$B_x = 900 \text{ N}$$

Next, find the length of diagonal BC.

$$\text{length of BC} = \sqrt{(6)^2 + (8)^2 + (4)^2}$$
$$= \sqrt{116}$$
$$= 10.77 \text{ m}$$

In the *x*-direction, cable BC has a length of 6 m and a force of 900 N. Since it has a total length of 10.77 m, the total force in BC is:

$$BC = \frac{10.77}{6} \times 900$$

$$\underline{BC = 1615 \text{ N } T}$$

Side View or *x-y* Plane

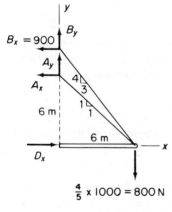

Figure 6-23

Go to the side view (Figure 6-23), where $B_x = 900$ N.

$$\Sigma M_D = 0$$

$$(900 \times 8) + (A_x \times 6) - (800 \times 6) = 0$$
$$A_x = -400 \text{ N} \leftarrow = +400 \text{ N} \rightarrow$$

Therefore, AC is in compression.

$$\text{length of } AC = \sqrt{(6)^2 + (6)^2} = 8.48 \text{ m}$$

$$\text{force in } AC = \frac{8.48}{6} \times 400$$

$$\underline{AC = 565 \text{ N } C}$$

$$\Sigma F_x = 0$$

$$D_x - B_x - A_x = 0$$
$$D_x - 900 - (-400) = 0$$
$$D_x = 500 \text{ N}$$

There are no other components at D, so the total load $\underline{DC = 500 \text{ N } C}$.

Example 6-9

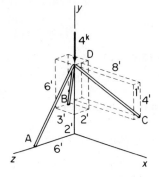

Figure 6-24

Solve for the compressive loads in members AD, BD, and CD in Figure 6-24. Points A, B, and C are ball and socket joints.

Calculate the diagonal length of each member.

$$AD = \sqrt{(8)^2 + (6)^2}$$
$$AD = 10 \text{ ft}$$
$$CD = \sqrt{(4)^2 + (8)^2 + (1)^2}$$
$$CD = 9 \text{ ft}$$
$$BD = \sqrt{(6)^2 + (3)^2 + (2)^2}$$
$$BD = 7 \text{ ft}$$

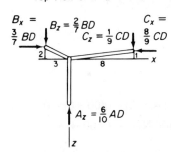

Top View or $x-z$ Plane

Figure 6-25

Draw the free-body diagram of the top view (Figure 6-25) and the front view (Figure 6-26) and label each component as a fraction of the total compressive force.

Taking moments about the same point in each diagram will give us simultaneous equations with two unknowns.

Take moments about C in the top view (Figure 6-25).

$$\Sigma M_C = 0$$

$$-\left(\frac{6}{10} AD \times 8\right) - \left(\frac{3}{7} BD \times 1\right)$$

$$+\left(\frac{2}{7} BD \times 11\right) = 0$$

$$4.8\,AD = 2.71\,BD$$
$$AD = 0.565\,BD \quad (1)$$

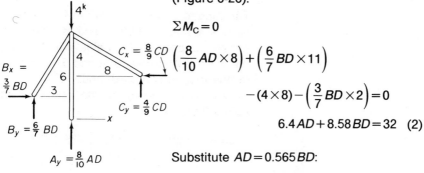

Front View or x–y Plane

Figure 6-26

Take moments about C in the front view (Figure 6-26).

$$\Sigma M_C = 0$$

$$\left(\frac{8}{10} AD \times 8\right) + \left(\frac{6}{7} BD \times 11\right)$$

$$-(4 \times 8) - \left(\frac{3}{7} BD \times 2\right) = 0$$

$$6.4 AD + 8.58 BD = 32 \quad (2)$$

Substitute $AD = 0.565 BD$:

$$3.62 BD + 8.58 BD = 32$$
$$\underline{BD = 2.63 \text{ kips } C}$$

From Equation (1):

$$AD = 0.565 \times 2.63$$
$$\underline{AD = 1.48 \text{ kips } C}$$

Take the sum of the forces in the x-direction in the top view.

$$\Sigma F_x = 0$$

$$\frac{3}{7} BD = \frac{8}{9} CD$$

$$CD = \frac{3 \times 9}{7 \times 8} \times 2.63$$

$$\underline{CD = 1.27 \text{ kips } C}$$

A check can be made by taking the sum of the forces in the y-direction in the front view.

$$\Sigma F_y = 0$$

$$\left(\frac{6}{7} \times 2.63\right) + \left(\frac{8}{10} \times 1.48\right)$$

$$+ \left(\frac{4}{9} \times 1.27\right) - 4 = 0$$

$$2.25 + 1.18 + 0.564 = 4$$

$$\underline{4 = 4} \quad (check)$$

Example 6-10

The hinged platform shown in Figure 6-27 carries a load of 600 N and is held in a horizontal position by cable AB. Neglect the weight of the platform and determine
(a) the load in cable AB, and
(b) the components of the reactions at points C and D.

$$\text{cable length} = \sqrt{(0.8)^2 + (1.2)^2 + (0.5)^2}$$
$$= 1.53 \text{m}$$

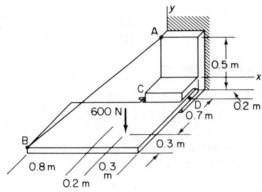

Figure 6-27

The components at B can be written as:

$$B_x = \frac{0.8}{1.53} AB = 0.524 AB$$

$$B_y = \frac{0.5}{1.53} AB = 0.328 AB$$

$$B_z = \frac{1.2}{1.53} AB = 0.786 AB$$

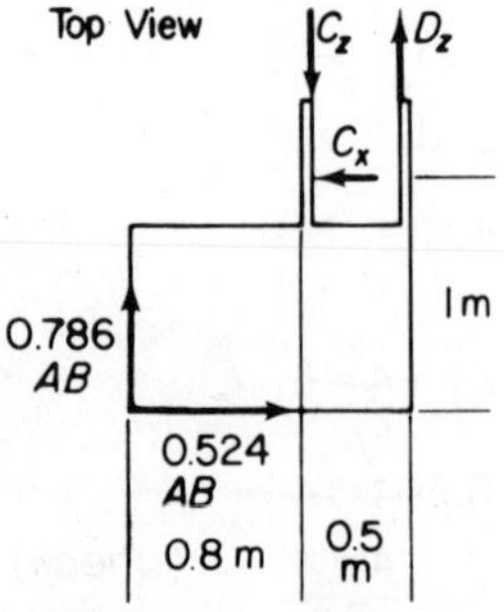

Draw top, front, and side views (Figures 6-28 to 6-30) showing these components of AB. Take moments about CD in the side view (Figure 6-30).

$$\Sigma M_{CD} = 0$$

$$-(0.328 AB \times 1) + (600 \times 0.7) = 0$$

$$AB = 1280 \text{N} \ T$$

Top View

Figure 6-28

Front View

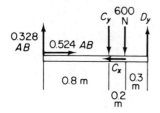

Figure 6-29

Side View

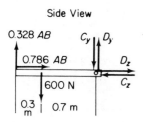

Figure 6-30

Now consider the front view (Figure 6-29).

$$\Sigma F_x = 0$$

$$C_x = 0.524 \times 1280$$

$$\underline{C_x = 671 \text{ N} \leftarrow}$$

$$\Sigma M_D = 0$$

$$(600 \times 0.3) + (C_y \times 0.5)$$

$$-(0.328 \times 1280 \times 1.3) = 0$$

$$\underline{C_y = 732 \text{ N} \downarrow}$$

$$\Sigma F_y = 0$$

$$0.328 \times 1280 + D_y = 732 + 600$$

$$\underline{D_y = 912 \text{ N} \uparrow}$$

The top view (Figure 6-28) will give us the remaining unknown components.

$$\Sigma M_D = 0$$

$$(C_z \times 0.5) + (0.524 \times 1280 \times 1)$$

$$-(0.786 \times 1280 \times 1.3) = 0$$

$$\underline{C_z = 1274 \text{ N} \swarrow}$$

$$\Sigma F_z = 0$$

$$D_z - 1274 + 0.786 \times 1280 = 0$$

$$\underline{D_z = 268 \text{ N} \nearrow}$$

PROBLEMS

6-1–6-6 Determine the magnitude and location of the resultant of the force systems shown in Figures 6-1 to 6-6.

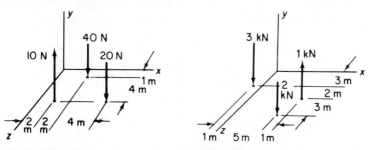

Figure P6-1

Figure P6-2

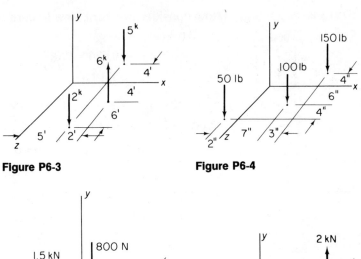

Figure P6-3   **Figure P6-4**

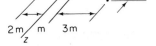

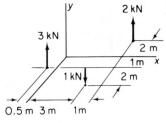

Figure P6-5   **Figure P6-6**

6-7 A 1000-lb platform that has a load of 6 kips on it is supported at points A, B, and C. Determine the force at each point of support (Figure P6-7).

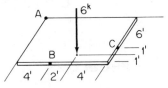

Figure P6-7

6-8 Determine the forces required at points A, B, and C to support the loads shown in Figure P6-8.

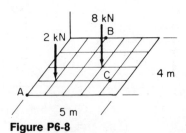

Figure P6-8

6-9 A platform supported by three ropes (shown in Figure P6-9) has a mass of 204 kg and carries a load of 4 kN. Calculate the tension in each rope.

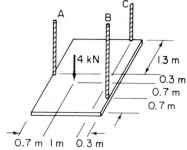

Figure P6-9

6-10 The three-wheeled cart in Figure P6-10 weighs 120 lb. If it is loaded as shown with a crate weighing 200 lb, find the load on each wheel.

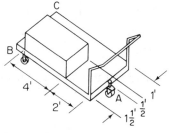

Figure P6-10

6-11 A piece of machinery weighing 5000 lb is lifted to its rooftop installation site by means of a crane. The vertical crane cables are attached at points A, B, and C. Determine the tension in each cable (Figure P6-11).

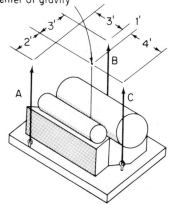

Figure P6-11

6-12 A carport roof is supported as shown in Figure P6-12. If the roof has a mass of 200 kg, determine the load on each support.

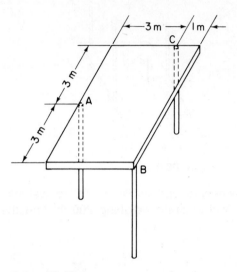

Figure P6-12

6-13 Determine the resultant of the forces shown in Figure P6-13. Find the coordinates.

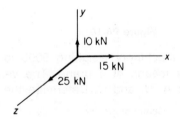

Figure P6-13

6-14 Determine the resultant of the forces shown in Figure P6-14.

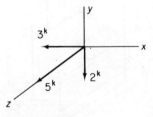

Figure P6-14

6-15   An anchor block is acted upon by the forces shown in Figure P6-15. Determine the resultant force.

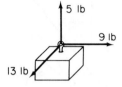

Figure P6-15

6-16   Determine the resultant of the forces shown in Figure P6-16.

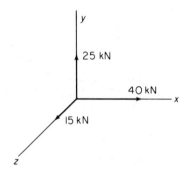

Figure P6-16

6-17   Determine the x, y, and z components of the force shown in Figure P6-17.

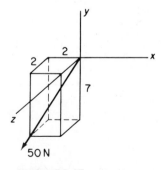

Figure P6-17

6-18 Determine the x, y, and z components of the force shown in Figure P6-18.

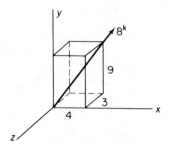

Figure P6-18

6-19 Determine the x, y, and z components of the force shown in Figure P6-19.

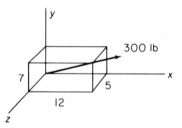

Figure P6-19

6-20 Determine the x, y, and z components of the force shown in Figure P6-20.

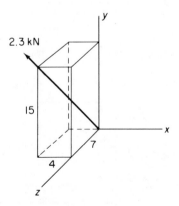

Figure P6-20

6-21 Leg AB of the derrick shown in Figure P6-21 is subject to a compressive load of 700 lb. Determine the x, y, and z component forces that it exerts on the ground at A.

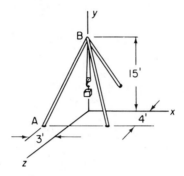

Figure P6-21

6-22 Determine the resultant of the forces shown in Figure P6-22.

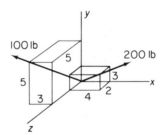

Figure P6-22

6-23 Determine the resultant of the forces shown in Figure P6-23.

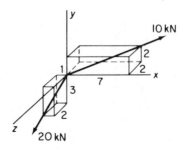

Figure P6-23

6-24   Determine the resultant of the forces shown in Figure P6-24.

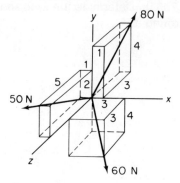

Figure P6-24

6-25   Determine the resultant of the forces shown in Figure P6-25.

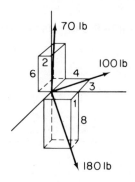

Figure P6-25

6-26   The shaft in Figure P6-26 receives an input torque due to the belt tensions shown and it transmits this torque to a machine at D. Determine the y and z components of the bearing reactions at B and C.

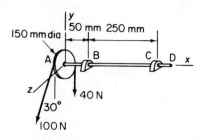

Figure P6-26

6-27 A counterweighted mechanism for tightening a belt at pulley E is shown in Figure P6-27. The counterweight weighs 20 lb. At an instant during which the belt is being loosened, the force at G is 5 lb, and the tensile force in AB is 14.5 lb. Determine the x and y components of the reactions at C and D.

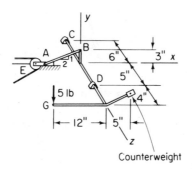

Figure P6-27

6-28 Determine the load in each member of the frame shown in Figure P6-28.

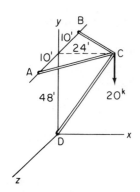

Figure P6-28

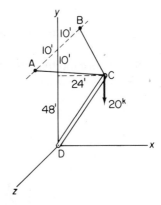

Figure P6-29

6-29 Determine the load in each member of the frame shown in Figure P6-29.

6-30 Determine the load in each member of the frame shown in Figure P6-30.

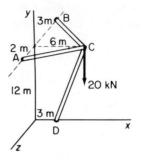

Figure P6-30

6-31 Determine the load in members AB, BC, and BD in Figure P6-31.

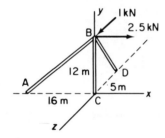

Figure P6-31

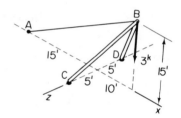

Figure P6-32

6-32 Determine the load in members AB, BC, and BD in Figure P6-32.

6-33 Determine the load in each of the three legs in Figure P6-33.

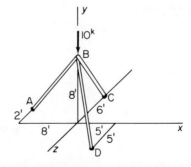

Figure P6-33

6-34 Determine the load in members AC, BC, and DC in Figure P6-34.

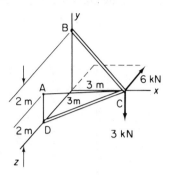

Figure P6-34

6-35 Antennae have the supporting cables shown in Figure P6-35. There are no other forces acting on the antennae other than the bottom socket connection. If cable AB is tightened to a tension of 300 lb, what is the tension in cables BC and BD.

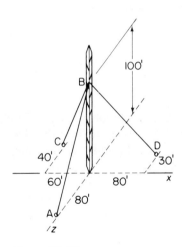

Figure P6-35

6-36 Determine the cable tensions if the tower has initial cable tensions of Problem 6-35 but subsequently experiences a wind force equivalent to 800 lb at B acting parallel to the z-axis and to the right. Assume no stretching of cable AB.

6-37 Pole AB is anchored to the side of a building as shown in Figure P6-37. Determine the load in members AB, CB, and DB.

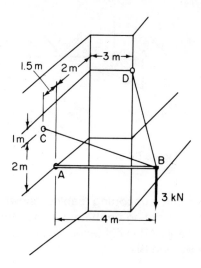

Figure P6-37

6-38 A tube handling bulk material in Figure P6-38 pivots at B and is controlled by adjustment of cables AD and DC. The combined weight of the tube and material is 1500 lb, and it acts through point D. Determine the cable tensions AD and CD.

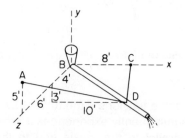

Figure P6-38

6-39 A sign with a mass of 120 kg has a wind force of 400 N on its face as shown (Figure P6-39). Members BE and FG are very short cable connectors and point D is a ball and socket connection. Determine (a) cable tensions AB, BC, and HK, and (b) reaction components at point D.

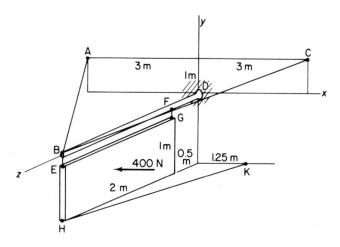

Figure P6-39

7

FRICTION

Much time and energy has been spent reducing unwanted friction in machines and engines. The internal combustion engine is a prime example of this attempt. Perhaps an equal amount of time has been spent trying to utilize friction. The design of tires and various braking systems is indicative of this effort. The understanding and use of friction is important in so many of our everyday activities—not to mention in equipment design—that the basic friction laws for dry surfaces will now be considered.

7-1   FRICTION LAWS FOR DRY SURFACES

The following discussion will concern only nonlubricated surfaces.

Motion or impending motion of two surfaces in contact causes a reaction force known as a *friction force, F.* This friction force is:

1. parallel to a flat surface or tangent to a curved surface;
2. opposite in direction to the motion or impending motion;

191

3. dependent on the force pressing the surfaces together;

4. generally independent of the area of surface of contact;

5. independent of velocity, except for extreme cases not to be considered here; and

6. dependent on the nature of the contacting surfaces.

Impending motion means that the object being considered is on the verge of moving; a small additional force would cause motion. An object said to be in impending motion is not moving and, therefore, is in static equilibrium. A free-body diagram showing all external forces can be drawn.

Friction forces can also exert their effects on an object in motion. This is a case of dynamic equilibrium where velocity and acceleration may be of concern. The dynamics portion of this book covers such a situation.

7-2   COEFFICIENTS OF FRICTION

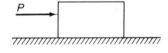

Figure 7-1

Consider the case of a block on a horizontal surface being pushed by a force *P* (Figure 7-1) so that the block has impending motion to the right. A free-body diagram of the block (Figure 7-2) would have the following forces.

Free–Body Diagram of Block

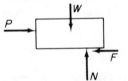

Figure 7-2

P—as shown

W—weight of the block

F—force of friction

N—normal force

Normal means *at right angles to*, so the normal force here is at right angles to the surface and is equal to the force pressing the two surfaces together. In this case, it is equal to the weight of the block.

There can be many cases in which the normal force is not equal to the weight, such as on an inclined surface, when the applied force *P* is not horizontal, and when other external forces are acting on the object. The use of a free-body diagram and equilibrium equations is necessary for these cases.

In order to describe the friction between two surfaces, we use the relationship between the friction force and the normal force. The friction force depends on the normal force and is always a fraction of the normal force. The friction force is therefore expressed as a fraction or portion of the normal force. This relationship is called the *coefficient of friction* and is represented by the Greek letter mu (μ).

$$\mu = \frac{F}{N} \qquad (7\text{-}1)$$

where F = friction force; N = normal force; and μ = coefficient of friction (a numerical value with no units).

There can be *static friction forces* (where there is impending motion) or *kinetic friction forces* (where the surfaces are moving with respect to each other). Since both static and kinetic friction forces exist, there are also static and kinetic coefficients of friction, μ_s and μ_k. Static friction will be our prime concern in this chapter, and we will simply use μ without the subscript "s."

The coefficient of friction can be determined for any two materials in contact but varies within a range of values. Because of this variation in values, the coefficient of friction of various surfaces is not given in a tabular form; rather, average values will be given for specific problems. A later example (Example 7-2) shows how the coefficient of friction can be determined experimentally.

7-3   ANGLE OF FRICTION

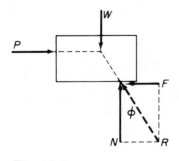

A second way in which friction is described is by the *angle of friction*. In Figure 7-3, the friction force F and the normal force N are combined in the resultant R. The friction angle ϕ (phi) is the angle between N and R. Considering the right-angle triangle in which ϕ is located, we can write:

$$\tan \phi = \frac{F}{N}$$

Figure 7-3

but the coefficient of friction, $\mu = \dfrac{F}{N}$; therefore:

$$\mu = \frac{F}{N} = \tan \phi$$

Friction can be expressed as a coefficient or as the tangent of the friction angle ϕ.

When drawing free-body diagrams in friction problems, you have the choice of showing F and N separately or in combination as R at some friction angle. The following examples will show that the choice of method is determined by which method offers the easiest solution.

Example 7-1

Figure 7-4

The 80-N force shown in Figure 7-4 causes impending motion to the right. Determine the static coefficient of friction μ.

Free—Body Diagram of Block

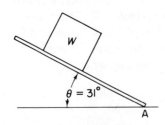

Figure 7-5

Since there are only two vertical forces in the free-body diagram of the block (Figure 7-5):

$$N = 400 \text{ N}\uparrow$$

Similarly, in the horizontal direction:

$$F = 80 \text{ N}\leftarrow$$

$$\mu = \frac{F}{N}$$

$$\mu = \frac{80}{400}$$

$$\underline{\mu = 0.2}$$

Example 7-2

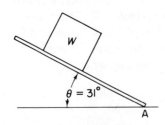

Figure 7-6

A block of weight W is placed on a plane that is pivoted at A (Figure 7-6). The plane is tilted until, at $\theta = 31°$, the block begins to slide. What is the coefficient of static friction μ between the block and the plane?

Free-Body Diagram of Block

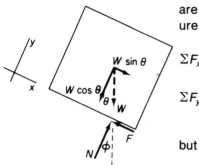

Figure 7-7

The weight is broken into components that are normal and parallel to the surface (Figure 7-7).

$$\Sigma F_x = 0$$

$$F = W \sin\theta$$

$$\Sigma F_y = 0$$

$$N = W \cos\theta$$

but

$$\mu = \frac{F}{N}$$

$$= \frac{W \sin\theta}{W \cos\theta}$$

By trigonometric definition:

$$\tan\theta = \frac{\sin\theta}{\cos\theta}$$

The weight W cancels, and we have:

$$\mu = \tan\theta$$

$$= \tan 31°$$

$$\underline{\mu = 0.6}$$

This is an experimental method of determining the static coefficient of friction by measuring the sloping plane angle that causes impending motion.

Example 7-3

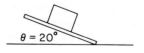

$\theta = 20°$

Figure 7-8

The block shown (Figure 7-8) has a force of gravity of 80 N and a coefficient of static friction of 0.5 between it and the sloped surface. Determine the force of friction acting on the block.

There is no mention of impending motion. We will find that the actual friction force cannot be found using $F = \mu N$.

Free–Body Diagram of Block

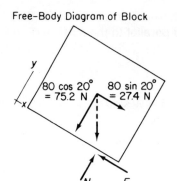

Figure 7-9

The force of gravity is resolved into components (Figure 7-9).

$$\Sigma F_y = 0$$

$$N = 75.2 \text{ lb}$$

The maximum available friction force would be:

$$F = \mu N$$
$$= 0.5 \times 75.2$$
$$F = 37.6 \text{ lb}$$

But from Figure 7-9, the actual friction force is found by:

$$\Sigma F_x = 0$$

$$\underline{F = 27.4 \text{ lb}}$$

We have an actual friction force and a maximum friction force since motion is not impending. At any time up to the point of impending motion, the actual friction force will only be large enough to keep the object in static equilibrium. At the point of impending motion, it will have reached its maximum value as given by $F = \mu N$.

The friction force in this case is 27.4 lb.

Example 7-4

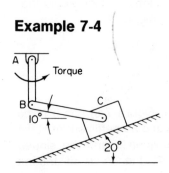

Figure 7-10

By means of a torque due to member AB, the block that has a mass of 20.4 kg is in a state of impending motion up the plane (Figure 7-10). Determine the compressive load in member BC if the coefficient of static friction between the block and the plane is 0.2.

The block has a force of gravity of $9.81 \times 20.4 = 200$ N. As shown in Figure 7-11, the force of friction F opposes the motion and is acting down the slope. If we attempt to

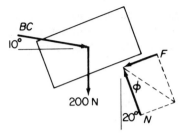

Figure 7-11

use this diagram, we will have to resolve some forces into components, and simultaneous equations will result.

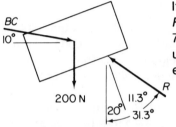

Figure 7-12

If F and N are combined into the resultant R, then there are only three forces (Figure 7-12), and a vector triangle solution can be used. The friction angle ϕ is found from the equation:

$$\mu = \tan \phi$$
$$0.2 = \tan \phi$$
$$\theta = 11.3°$$

Construct the vector triangle (Figure 7-13) by drawing first the 200 N vector, then vector BC, and finally vector R, which closes the triangle.

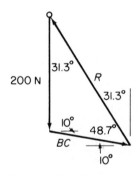

Figure 7-13

Applying the sine law, we get:

$$\frac{BC}{\sin 31.3°} = \frac{200}{\sin 48.7}$$
$$BC = 200 \times \frac{0.52}{.751}$$
$$\underline{BC = 138 \text{ N } C}$$

Example 7-5

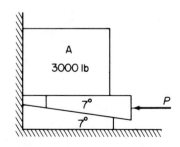

Figure 7-14

A 3000-lb weight is raised by two 7° wedges as shown in Figure 7-14. The coefficient of static friction is 0.23 for all surfaces. Determine the minimum force P.

Since the friction force and normal force will be combined into a resultant in all free-body diagrams, the friction angle is calculated first.

$$\mu = \tan \phi$$
$$0.23 = \tan \phi$$
$$\phi = 13°$$

Free–Body Diagram of A

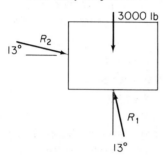

Figure 7-15

Since the only known force is the 3000-lb weight, we start with a free-body diagram of weight A (Figure 7-15). The corresponding vector triangle is shown in Figure 7-16.

Only the value of R_1 is required since this will allow us to draw a free-body diagram of the top wedge and to solve for P.

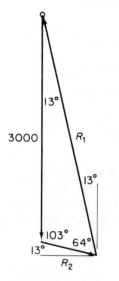

Figure 7-16

$$\frac{R_1}{\sin 103°} = \frac{3000}{\sin 64°}$$
$$R_1 = 3000 \times \frac{\sin 77°}{\sin 64°}$$
$$= 3000 \times \frac{0.975}{0.899}$$
$$R_1 = 3260 \text{ lb}$$

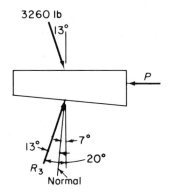

Figure 7-17

The value of R_1 acting on block A (Figure 7-15) having been found, it is now shown as an equal but opposite in direction force acting on the top wedge (Figure 7-17).

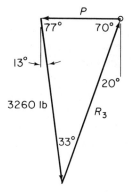

Figure 7-18

From the vector triangle (Figure 7-18):

$$\frac{P}{\sin 33°} = \frac{3260}{\sin 70°}$$

$$P = 3260 \times \frac{0.545}{0.94}$$

$$\underline{P = 1890 \text{ lb} \leftarrow}$$

7-4 BELT FRICTION

The two main assumptions in this section are:

1. A rope, cable, or flat belt is used. (Using a notched belt or a V-belt would require further analysis.)
2. Motion is impending—in the manner of a rope wound around a fixed cylinder so that the rope is starting to slip.

Consider the simplified case of a rope passed over a fixed cylindrical beam and a force P causing impending motion of the

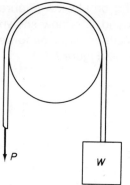

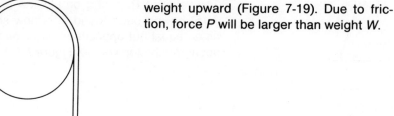

weight upward (Figure 7-19). Due to friction, force P will be larger than weight W.

Figure 7-19

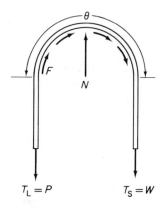

Figure 7-20

The rope has two different tensions, a large tension (T_L) and a small tension (T_S). These are shown in a free-body diagram of the portion of rope passing over the cylinder (Figure 7-20). Notice that the friction force is acting in a direction opposite to that of the impending motion of the cylinder with respect to the rope. A free-body diagram of the rope shows the force of friction on the rope due to the cylinder. There is impending motion of the cylinder with respect to the rope, so we show the force of the cylinder on the rope.

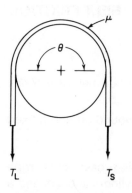

Figure 7-21

For impending slipping (Figure 7-21), the difference between T_L and T_S depends on:

1. the coefficient of friction μ; and
2. the angle of contact θ.

This is expressed in the equation:

$$\log_{10} \frac{T_L}{T_S} = 0.434\,\mu\theta \qquad (7\text{-}2)$$

where T_L = large tension; T_S = small tension; μ = static coefficient of friction; and θ = angle of contact in radians.

An alternate equation that uses natural logs is

$$\ln \frac{T_L}{T_S} = \mu\theta \tag{7-3}$$

or

$$\frac{T_L}{T_S} = e^{\mu\theta} \tag{7-4}$$

where e = 2.718, the Naperian constant; and θ = angle of contact in radians.

Your choice of which belt friction equation to use will depend on which level of mathematics you are most familiar with. Equation 7-3 is generally preferred. Choose one of the equations and use it consistently to avoid confusion.

Example 7-6

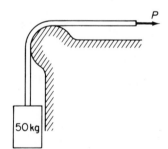

Figure 7-22

The 50-kg mass in Figure 7-22 has impending motion upward when P = 750 N. Determine the coefficient of friction between the rope and the cylinder.

Since P must overcome friction and lift the 50-kg:

$$T_L = P = 750 \text{ N}$$

and

$$T_S = 9.81 \times 50 = 490 \text{ N}$$

$$\theta = 90° = \frac{\pi}{2} \text{ rad}$$

$$\log_{10}\left(\frac{T_L}{T_S}\right) = 0.434\,\mu\theta$$

$$\log_{10}\left(\frac{750}{490}\right) = 0.434\,\mu \times \frac{\pi}{2}$$

$$\log_{10} 1.53 = 0.682\,\mu$$

But, $\log_{10} 1.53 = 0.185$ (by reading 1.53 on the D-scale and 0.185 on the L-scale).

$$0.185 = 0.682 \mu$$
$$\underline{\mu = 0.27}$$

Example 7-7

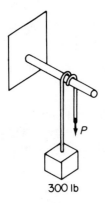

300 lb

Figure 7-23

Determine the minimum force P required to hold the weight of 300 lb in Figure 7-23. There are $1\frac{1}{2}$ turns of rope about the horizontal rod, and the coefficient of static friction is 0.2.

Since motion of the weight is impending downward, friction aids P and:

$$T_L = 300$$
$$T_S = P$$
$$\mu = 0.2$$
$$\theta = 1.5 \times 2\pi$$

Using the formula with logs to the base 10, we get:

$$\log_{10}\left(\frac{T_L}{T_S}\right) = 0.434 \mu\theta$$

$$\log_{10}\left(\frac{300}{P}\right) = 0.434 \times 0.2 \times 1.5 \times 2\pi$$

$$\log_{10}\left(\frac{300}{P}\right) = 0.818$$

Read 0.818 on the L-scale and 6.58 on the D-scale; that is, $\log_{10} 6.58 = 0.818$. Therefore:

$$\frac{300}{P} = 6.58$$

$$\underline{P = 45.6 \text{ lb}\downarrow}$$

The alternate method using natural logs is

as follows:

$$\ln\left(\frac{T_L}{T_S}\right)=\mu\theta$$

where $\theta=1.5\times2\pi$ radians,

$$\ln\left(\frac{300}{P}\right)=0.2\times1.5\times2\pi,$$

$$\ln\left(\frac{300}{P}\right)=1.88$$

Read 1.88 on the D-scale and 6.58 on the LL3-scale; that is, ln 6.58=1.88.

$$\frac{300}{P}=6.58$$

$$\underline{P=45.6\ \text{lb}\downarrow}$$

Example 7-8

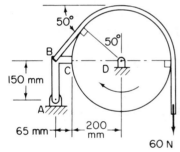

Figure 7-24

Lever AB and a belt work together to brake a wheel that is turning clockwise (Figure 7-24). The coefficient of kinetic friction for all surfaces is 0.3. What is the normal force between lever AC and the wheel if the belt is pulled with a force of 60 N?

The friction of the wheel on the belt aids the 60-N force; the small tension $T_S=60$ N.

$$\ln\left(\frac{T_L}{T_S}\right)=\mu\theta$$

where $\theta=\dfrac{140}{360}\times2\pi=2.44$ rad

$$\ln\left(\frac{T_L}{60}\right)=0.3\times2.44$$

$$\ln\left(\frac{T_L}{60}\right)=0.732$$

$$\left(\frac{T_L}{60}\right)=2.08$$

$$T_L=125\ \text{N}$$

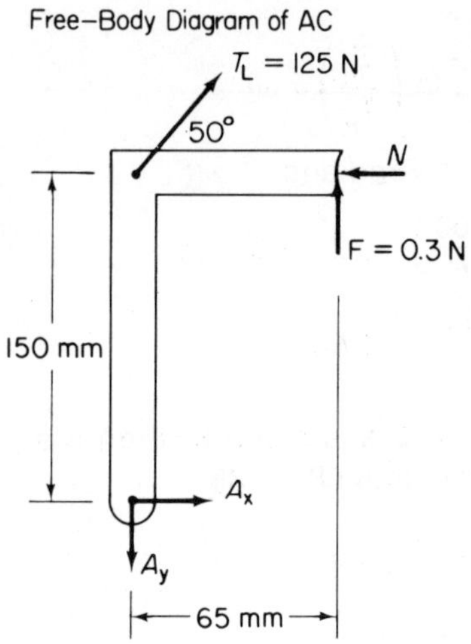

Free–Body Diagram of AC

$T_L = 125$ N

50°

N

$F = 0.3$ N

150 mm

A_x

A_y

|← 65 mm →|

Figure 7-25

Use a free-body diagram of member AC (Figure 7-25), where

$$\mu_k = \frac{F}{N}$$

$$F = 0.3N$$

$$\Sigma M_A = 0$$

$$(N \times 150) + (0.3N \times 65)$$

$$-(125 \cos 50° \times 150) = 0$$

$$169.5N = 12,050$$

$$\underline{N = 71.1 \text{ N}}$$

PROBLEMS

7-1 A horizontal force of 15 lb is required to start a 60-lb block sliding along a horizontal surface. What is the coefficient of static friction?

7-2 Determine the mass of a block if a force of 15 N is required to start it sliding on a horizontal surface ($\mu = 0.40$).

7-3 Determine the value of P for impending motion (a) down the slope, and (b) up the slope in Figure P7-3).

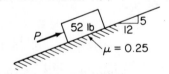

52 lb

5

12

$\mu = 0.25$

P

Figure P7-3

7-4 Determine the force P for impending motion up the plane shown in Figure P7-4.

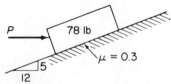

P

78 lb

5

12

$\mu = 0.3$

Figure P7-4

7-5 The block shown in Figure P7-5 has a mass of 34.7 kg. Determine the horizontal force P for impending motion down the plane.

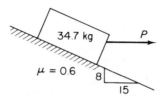

Figure P7-5

7-6 A 30-lb block has impending motion down a slope inclined 25° from the horizontal. What is the coefficient of static friction?

7-7 Will a mass of 10 kg slide—due to its own weight—down a slope inclined 30° to the horizontal if the coefficient of static friction is 0.6?

7-8 Determine the minimum force P that can hold the 13.5-lb weight shown in Figure P7-8.

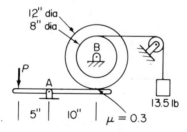

Figure P7-8

7-9 A canned-goods dispenser has a vertical column of 10 cans, each having a mass of 1 kg (Figure P7-9). The cans fit loosely in the vertical slot, and the coefficient of friction is 0.2 for all surfaces. What force P is required to pull the bottom can out?

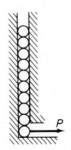

Figure P7-9

7-10 Determine the force P that will cause the impending motion of block A. What is the friction force acting on the bottom of block B (Figure P7-10)?

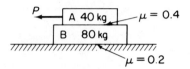

Figure P7-10

7-11 Determine the minimum force P that will cause impending motion of the block shown in Figure P7-11. Will the block tip or slide?

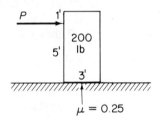

Figure P7-11

7-12 If force P in Figure P7-11 is now applied to the middle of the left side of the block and is acting down to the right at 30° to the horizontal, find the values of P for both tipping and sliding.

7-13 Block A in Figure P7-13 has a mass of 4 kg and is on the verge of tipping as it begins to slide due to force P. Determine the coefficient of static friction between the block and the horizontal surface.

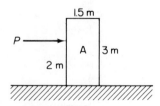

Figure P7-13

7-14 Each crate on the sloping chute in Figure P7-14 weighs 39 lb. The coefficient of static friction is 0.3. A spring-loaded roller at B applies a constant force of 54 lb. Determine the maximum number of crates that can be held on the chute before the bottom crate is pushed by the top crates onto the rollers at A.

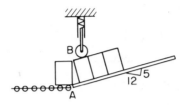

Figure P7-14

7-15 Wedge A is used to raise the 3000-lb block B in Figure P7-15. (The coefficient of static friction for all surfaces is 0.30.) Determine P for impending motion of B upward.

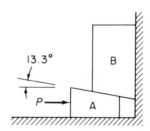

Figure P7-15

7-16 Determine the minimum force P that will pull wedge A to the left in Figure P7-16.

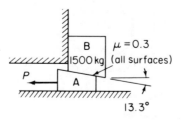

Figure P7-16

7-17 A 10° wedge is used to move the 50-kg mass in Figure P7-17. (The coefficient of static friction is 0.3 for all surfaces.) Determine force *P* for impending motion.

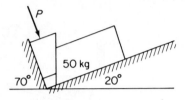

Figure P7-17

7-18 Determine the force *P* necessary to cause impending motion of the 40-lb block up the slope in Figure P7-18. The coefficient of static friction for all surfaces is 0.1.

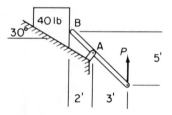

Figure P7-18

7-19 Neglect the weight of A and determine the vertical force of *P* necessary to cause impending motion of block B to the right in Figure P7-19. The coefficient of friction for all surfaces is 0.1.

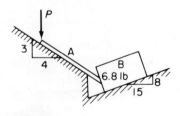

Figure P7-19

7-20 Slider A is moved to the right against a compressive spring force of 100 N by means of the eccentric lever B and force P (Figure P7-20). If the coefficient of static friction is 0.2, determine force P. (Neglect the weight of A.)

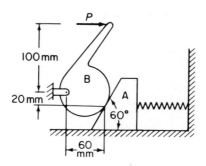

Figure P7-20

7-21 Determine the force P that will cause impending motion in Figure P7-21. The coefficient of static friction for all surfaces is 0.25.

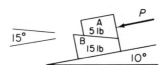

Figure P7-21

7-22 Determine the torque T required at D to cause impending motion of BD clockwise (Figure P7-22). The coefficient of friction at B is 0.3.

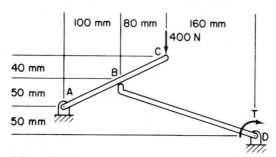

Figure P7-22

7-23 Determine the normal forces and friction forces acting on member AC (Figure P7-23) if it has impending motion when $P=700$ N. The coefficient of friction at B is 0.3. Determine μ at A.

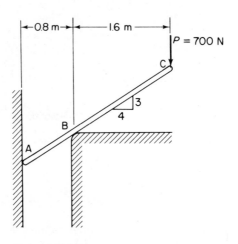

Figure P7-23

7-24 A shaft, 4 in. in diameter and weighing 100 lb, has impending motion counterclockwise due to an applied torque of 450 lb-in. (The coefficient of static friction for all surfaces is 0.25.) Determine the reactions at A and B on the shaft (Figure P7-24).

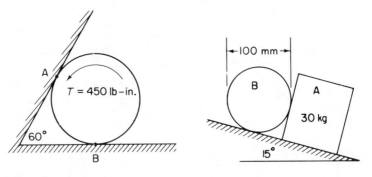

Figure P7-24   **Figure P7-25**

7-25 The coefficient of static friction is 0.45 for all surfaces in Figure P7-25. Determine the mass of cylinder B that will cause impending motion of both A and B down the plane.

7-26 Will the 40-kg cylinder in Figure P7-26 slip at either A or B? If it does slip, determine the actual friction forces at each of these points. The coefficient of static friction is 0.25 at A and 0.18 at B.

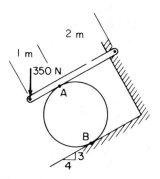

Figure P7-26

7-27 A 200-lb load is applied to the beam system in Figure P7-27, causing a tension of 160 lb in spring DC. Will slipping occur at A if the coefficient of static friction at this point is 0.2? What is the actual value of the friction force at A? What is the minimum allowable coefficient of static friction that will prevent motion at point A?

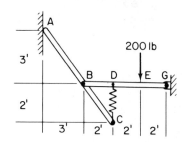

Figure P7-27

7-28 Calculate the force of friction acting upon block A of the system in Figure P7-28. The coefficient of static friction for all surfaces is 0.3.

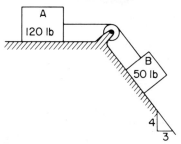

Figure P7-28

7-29 What weight of block B is required to produce impending motion of block A to the right in Figure P7-28?

7-30 Blocks A and B have masses of 60 kg and 50 kg, respectively (Figure P7-30). Calculate the coefficient of friction between block B and the sloped surface if there is impending motion of B down the slope.

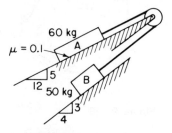

Figure P7-30

7-31 Determine the force T that will produce impending motion upward of the 500-lb weight in Figure P7-31. (Assume a coefficient of friction of 0.23.) Also find the minimum force T required to hold the 500-lb weight.

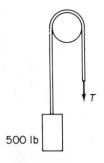

Figure P7-31

7-32 Determine the mass that can be lifted by a rope with 3/4 of a turn around a fixed horizontal shaft ($\mu = 0.3$) if a force of 8 kN is applied to its other end.

7-33 When mass B is 340 kg, it produces impending motion of block A (Figure P7-33). What is the coefficient of static friction between the rope and the fixed shaft?

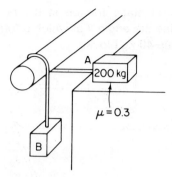

Figure P7-33

7-34 How many turns of rope around a horizontal fixed shaft are required if a force of 80 lb on the end of the rope is to hold a weight of 720 lb? Assume the coefficient of static friction to be 0.35.

7-35 The coefficient of static friction for all surfaces in Figure P7-35 is 0.16. Determine weight B that will cause impending motion of block A to the right.

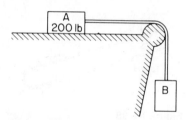

Figure P7-35

7-36 Block A in Figure P7-36 has a mass of 500 kg and is in a state of impending motion down the slope. Determine the mass of block B.

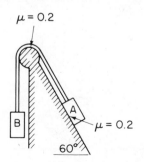

Figure P7-36

7-37 The coefficient of static friction is 0.4 between all surfaces in Figure P7-37. Determine the angle θ at which a force of 375 N will cause impending motion of the 40-kg block.

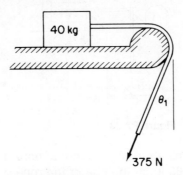

Figure P7-37

7-38 The tension in the top portion of the conveyor belt in Figure P7-38 is 1600 N. The coefficient of static friction between the belt and drum A is 0.3. What is the maximum torque that drum A may deliver before slipping occurs?

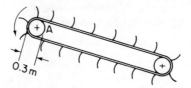

Figure P7-38

7-39 A rotating wheel is braked by the rope shown in Figure P7-39. (The coefficient of kinetic friction is 0.12.) Determine the decelerating torque applied to the wheel if $T = 15$ lb.

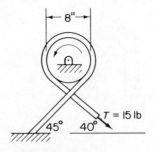

Figure P7-39

7-40 Determine the minimum force *P* necessary for wheel A to have impending motion clockwise due to an applied torque of 24 N·m clockwise. The coefficient of static friction between the belt and wheel is 0.3 (Figure P7-40).

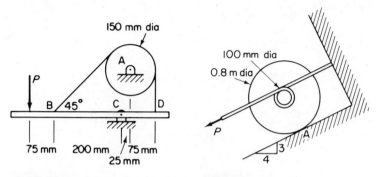

Figure P7-40   **Figure P7-41**

7-41 Cylinder A with a mass of 25 kg has a fixed hub about which there is one turn of rope as shown in Figure P7-41. (The coefficient of friction is 0.3 for all surfaces.) Determine the minimum force *P* that will prevent slipping of the rope on the hub. Will slipping occur at point A?

7-42 The structural shape in Figure P7-42 pivots at point A as it is lowered by slipping on the rope. The structural shape weighs 300 lb, and this weight may be assumed at the center of gravity as shown. If the coefficient of static friction is 0.38, determine the tension *T* if there is to be impending motion of the structural shape downward.

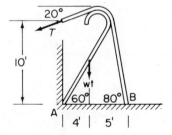

Figure P7-42

8

CENTROIDS AND
CENTER OF GRAVITY

8-1 INTRODUCTION

In section 4-7, we covered nonuniform loading of beams. When calculating the beam reactions, we were able to show the entire nonuniform load as acting at its center of gravity. The problem became one of determining the amount of the nonuniform load and locating the center of gravity. Center of gravity (C of G) or center of mass refers to masses or weights and can be thought of as the single point at which the weight could be held and be in balance in all directions.

If the weight or object were homogeneous, the center of gravity and centroid would coincide. In the case of a hammer with a wooden handle, its center of gravity would be close to the heavy metal end. The *centroid*, which is found by ignoring weight and considering only volume, would be closer to the middle of the handle. Due to varying densities, the center of gravity and centroid do not always coincide. Centroid usually refers to lines, areas, and volumes. It is also the central or balance point of the line, area, or volume.

The two main areas in which you will use centroids are in fluid mechanics and stress analysis. An example of centroids used in stress analysis is a beam that is caused to bend due to a load. The beam has a cross-sectional area, and locating the centroid of this area is one of the first steps in solving for the stress or deflection of the beam. In fluid mechanics, a vertical wall, such as a dam, is subjected to a water pressure that varies from zero at the top to a maximum at the bottom. Determining the centroid of this pressure distribution is necessary.

8-2   CENTROIDS OF SIMPLE AREAS

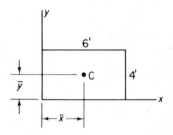

Figure 8-1

To see how centroids of areas are located, consider the area shown in Figure 8-1. Although the answer may be obvious to you as $\bar{x}=3$ ft and $\bar{y}=2$ ft, this example is used here for illustration purposes only. An odd-shaped area would result in more involved calculations.

The first step of either the calculus method or the method described below is division of the area into strips of equal width.

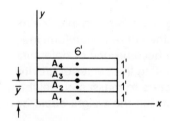

Figure 8-2

Divide the area into horizontal strips, arbitrarily selecting 1 ft as the width (Figure 8-2). A vector will represent the area of each strip. The vector equals the strip area and acts in a direction perpendicular to the strip at the center, or centroid, of the strip. Each vector or strip area has a moment about the x-axis. The sum of the moments of all strip areas is equal to the total area multiplied by the centroid distance $\bar{y}$.

$$A_1\, y_1 + A_2\, y_2 + A_3\, y_3 + A_4\, y_4 = A\bar{y}$$

$$(6\times0.5)+(6\times1.5)+(6\times2.5)$$
$$+(6\times3.5)=24\bar{y}$$
$$48=24\bar{y}$$
$$\bar{y}=2 \text{ ft}$$

The centroid distance $\bar{x}$ is found by dividing the area into vertical strips and taking moments about the y-axis (Figure 8-3). Multiply each area by its centroid distance.

$$A_1x_1 + A_2x_2 + A_3x_3 + A_4x_4 + A_5x_5 + A_6x_6 = A\bar{x}$$

$$(4 \times 0.5) + (4 \times 1.5) + (4 \times 2.5)$$

$$+ (4 \times 3.5) + (4 \times 4.5) + (4 \times 5.5)$$

$$= 24\bar{x}$$
$$\bar{x} = 3 \text{ ft}$$

To restate the above principle: *the moment of an area equals the algebraic sum of the moments of its component areas.* We will use this principle in locating the centroid of composite areas (Section 8-3).

For the simpler geometric areas, there is a faster method of locating the centroid, C, than the one set forth above. First observe that a line drawn through the centroids of the horizontal strips (Figure 8-2) and a line drawn through the centroids of the vertical strips (Figure 8-3) intersect at point C, the centroid of the area. Apply the same reasoning to the triangle in Figure 8-4 by taking elemental strips parallel to the base of the triangle and parallel to the left side. The loci of the midpoints of the strips intersect at C, the centroid of the triangle. The indicated proportions of the height h are found to exist. The height is always measured perpendicular to the base.

The centroids of some of the more common areas are shown in Table 8-1.

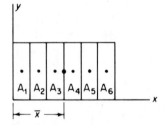

Figure 8-3

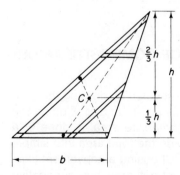

Figure 8-4

Table 8-1 — Centroids of Areas

SHAPE		$\bar{X}$	$\bar{Y}$	AREA
1. Triangle			$\dfrac{h}{3}$	$\dfrac{1}{2}bh$
2. Semicircle		0	$\dfrac{4r}{3\pi}$	$\dfrac{\pi r^2}{2}$
3. Quarter circle		$\dfrac{4r}{3\pi}$	$\dfrac{4r}{3\pi}$	$\dfrac{\pi r^2}{4}$

8-3 CENTROIDS OF COMPOSITE AREAS

We have seen how the centroid of an area is located by dividing the area into elemental strips and adding the moments about an axis. Finding the centroid of a more complex area, i.e., a *composite area*, is done in a similar manner; the difference is that we divide the total area into simple geometric areas that have known centroids. Choosing a convenient axis, we sum the moments of these areas. A cut-out area has a negative moment.

Example 8-1

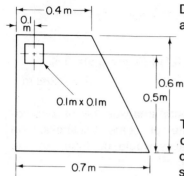

Figure 8-5

Determine the centroid of the composite area shown in Figure 8-5.

The first step is to choose and label all component areas as in Figure 8-6. The centroid distances for each area are shown. Any cut-out area—such as the square— is considered to have a negative moment. Calculate each area.

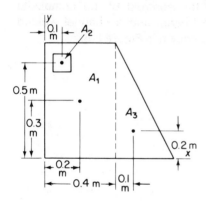

Figure 8-6

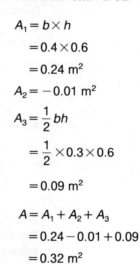

$$A_1 = b \times h$$
$$= 0.4 \times 0.6$$
$$= 0.24 \text{ m}^2$$
$$A_2 = -0.01 \text{ m}^2$$
$$A_3 = \frac{1}{2} bh$$
$$= \frac{1}{2} \times 0.3 \times 0.6$$
$$= 0.09 \text{ m}^2$$
$$A = A_1 + A_2 + A_3$$
$$= 0.24 - 0.01 + 0.09$$
$$= 0.32 \text{ m}^2$$

Take moments about the x-axis to find $\bar{y}$ (Figure 8-7).

$$A_1 y_1 + A_2 y_2 + A_3 y_3 = A\bar{y}$$

$$(0.24 \times 0.3)$$
$$+ (-0.01 \times 0.5) + (0.09 \times 0.2) = 0.32\bar{y}$$
$$0.072 - 0.005 + 0.018 = 0.32\bar{y}$$
$$\underline{\bar{y} = 0.266 \text{ m}}$$

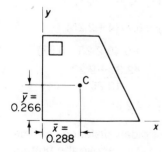

Figure 8-7

Moments about the y-axis will give $\bar{x}$ (Figure 8-7).

$$A_1 x_1 + A_2 x_2 + A_3 x_3 = A\bar{x}$$

(0.24×0.2)
$$+(-0.01 \times 0.1)+(0.09 \times 0.5)=0.32\bar{x}$$
$$\bar{x} = 0.288 \text{ m}$$

If this composite area was that of a homogeneous plate of some thickness, the center of gravity would be found by the same method.

Example 8-2

Determine the centroid of the composite area of an I-beam and a channel welded together as shown in Figure 8-8.

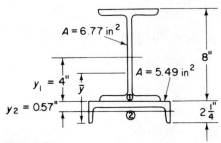

Figure 8-8

The centroid distances shown on the left of the diagram and areas are available from tables in various handbooks or stress texts.

$$A\bar{y} = A_1 y_1 + A_2 y_2$$

$$(6.77 + 5.49)\bar{y} = 6.77(4 + 2.25)$$
$$+5.49(2.25 - 0.57)$$
$$\bar{y} = \frac{42.31 + 9.22}{12.26}$$
$$\bar{y} = 4.2 \text{ in}$$

The vertical centroidal distance of the composite area is 4.2 in. above the bottom of the area.

8-4   CENTROIDS OF LINES

Centroids are not restricted to areas. A thin material may be in the form of a *complex profile* or *cross section.* The centroid of such a cross section is the *centroid of lines.* The situation is parallel to that encountered when we deal with areas. The total line length multiplied by its centroid distance from an axis is equal to the sum of each segment line length multiplied by its centroidal distance. Using $\bar{x}$ and $\bar{y}$ as centroidal distances and L for length, we have:

$$L\bar{x} = L_1 x_1 + L_2 x_2 + L_3 x_3 + \cdots$$

$$L\bar{y} = L_1 y_1 + L_2 y_2 + L_3 y_3 + \cdots$$

Some common variations of line centroids are given in Table 8-2.

Table 8-2 — **Centroids of Lines**

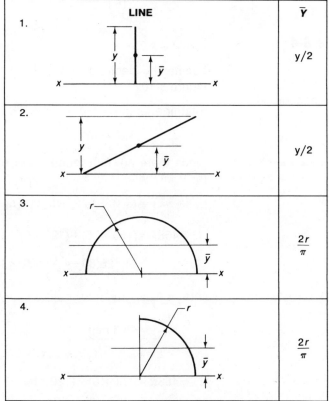

	LINE	$\bar{Y}$
1.		$y/2$
2.		$y/2$
3.		$\dfrac{2r}{\pi}$
4.		$\dfrac{2r}{\pi}$

Example 8-3

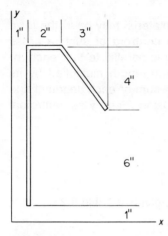

Figure 8-9

Locate the centroid of the line shown in Figure 8-9.

$$\text{total line length} = 10 + 2 + 5 = 17 \text{ in.}$$

There are three simple lengths or shapes. With the x-axis as our base, we solve for $\bar{y}$.

$$17\bar{y} = (10 \times 6) + (2 \times 11) + (5 \times 9)$$

$$\bar{y} = \frac{60 + 22 + 45}{17}$$

$$\underline{y = 7.47 \text{ in.}}$$

$$17\bar{x} = (10 \times 1) + (2 \times 2) + (5 \times 4.5)$$

$$\bar{x} = \frac{10 + 4 + 22.5}{17}$$

$$\underline{\bar{x} = 2.15 \text{ in.}}$$

Example 8-4

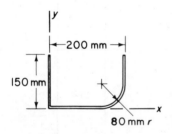

Figure 8-10

Locate the centroid of the line shown in Figure 8-10.

$$\text{total line length} = 150 + 120 + \frac{2\pi \times 80}{4} + 70$$

$$= 466 \text{ mm}$$

Since there are four lengths, our equation for $\bar{y}$ will be:

$$L\bar{y} = L_1 y_1 + L_2 y_2 + L_3 y_3 + L_4 y_4$$

$$466\bar{y} = (150 \times 75) + (120 \times 0)$$

$$+ 126(80 - \frac{2r}{\pi}) + 70 \times 115$$

$$\bar{y} = \frac{11,250 + 0 + 3663 + 8,050}{466}$$

$$\underline{\bar{y} = 49.3 \text{ mm}}$$

$$L\bar{x} = L_1 x_1 + L_2 x_2 + L_3 x_3 + L_4 x_4$$

$$466\bar{x} = (150 \times 0) + (120 \times 60)$$

$$+ 126(120 + \frac{2r}{\pi}) + 70 \times 200$$

$$\bar{x} = \frac{0 + 7{,}200 + 21{,}537 + 14{,}000}{466}$$

$$\bar{x} = 91.7 \text{ mm}$$

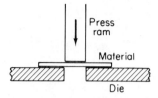

Press ram

Material

Die

Figure 8-11

Punching or shearing an area from a sheet by using a punch press and die is an operation that may require determination of the centroid of lines (Figure 8-11). The cutting forces around the outside of the shape should be symmetrically distributed with respect to the press ram. This is known as the *center of pressure* of the die and is simply the centroid of the outside lines of the shape.

Example 8-5

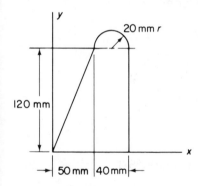

Figure 8-12

Determine the center of pressure of the blank in Figure 8-12.

We are to find the centroid of the lines that form the shape or blank shown.

There are four simple lines giving us a total line length of

$$90 + 130 + \frac{2\pi \times 20}{2} + 120 = 403 \text{ mm}$$

The distance from the diameter to the centroid of the semicircle is $2r/\pi$.

$$403\bar{y} = (90 \times 0) + (130 \times 60)$$

$$+ \left[\left(\frac{2 \times 20}{\pi} + 120\right)62.8\right] + (120 \times 60)$$

$$\bar{y} = \frac{0 + 7800 + 8336 + 7200}{403}$$

$$\bar{y} = 57.9 \text{ mm}$$

$$403\bar{x} = (90 \times 45) + (130 \times 25)$$

$$+ (62.8 \times 70) + (120 \times 90)$$

$$\bar{x} = 55.8 \text{ mm}$$

PROBLEMS

8-1   Determine $\bar{y}$ for the area shown in Figure P8-1.

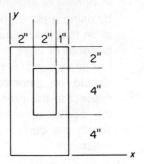

Figure P8-1

8-2   Determine $\bar{x}$ for the area shown in Figure P8-2.

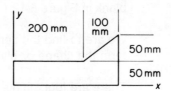

Figure P8-2

8-3   Locate the centroid of the area shown in Figure P8-3.

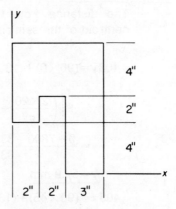

Figure P8-3

8-4 Locate the centroid of the area shown in Figure P8-4.

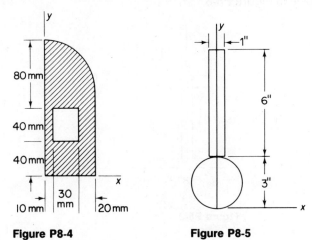

Figure P8-4   **Figure P8-5**

8-5 Locate the centroid of the area shown in Figure P8-5.

8-6 Locate the centroid of the channel cross section shown in Figure P8-6.

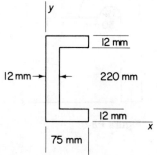

Figure P8-6

8-7 Locate the centroid of the cross-sectional area shown in Figure P8-7.

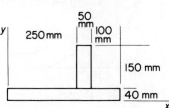

Figure P8-7

8-8   Locate the centroid of the cross-sectional area of the fabricated I-beam shown in Figure P8-8.

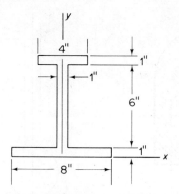

Figure P8-8

8-9   Locate the centroid of the area shown in Figure P8-9.

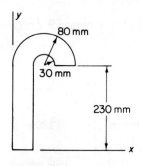

Figure P8-9

8-10   Locate the centroid of the area shown in Figure P8-10.

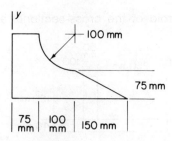

Figure P8-10

8-11 Locate the centroid of the area shown in Figure P8-11.

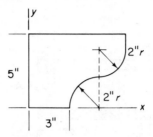

Figure P8-11

8-12 Determine $\bar{y}$ for the area shown in Figure P8-12.

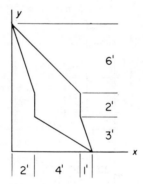

Figure P8-12

8-13 Determine $\bar{y}$ for the combined area of a channel and angle joined as shown in Figure P8-13. The individual areas and centroidal distances are given.

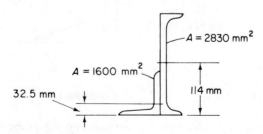

Figure P8-13

8-14 Locate the centroid of the cross-sectional area of the beam shown in Figure P8-14.

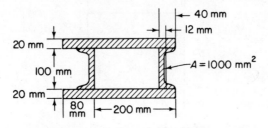

Figure P8-14

8-15 Locate the centroid of the extruded shape shown in Figure P8-15.

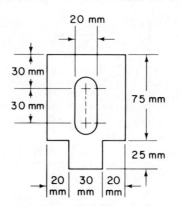

Figure P8-15

8-16 Locate the centroid of the line shown in Figure P8-16.

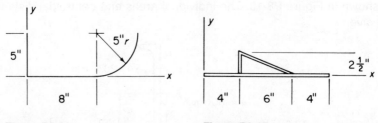

Figure P8-16   **Figure P8-17**

8-17 Sheet material is formed and welded into the cross section shown in Figure P8-17. Locate the centroid.

8-18 Locate the centroid of a rod bent to the shape shown in Figure P8-18.

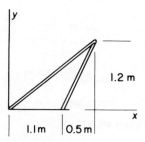

Figure P8-18

8-19 The line profile of the cross section formed by fabricated thin sheet metal is shown in Figure P8-19. Locate the centroid.

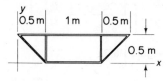

Figure P8-19

8-20 The cross-sectional area shown in Figure P8-20 is to be punched from sheet metal with a punch press. Determine the center of pressure.

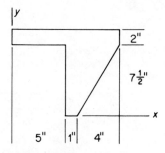

Figure P8-20

8-21 A thin wire is bent to form the shape shown in Figure P8-21. Locate the centroid.

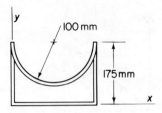

Figure P8-21

8-22 Locate the centroid of the line that is the perimeter of the area shown in Figure P8-22.

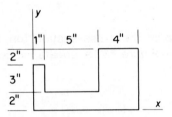

Figure P8-22

9

MOMENT OF INERTIA

9-1 MOMENT OF INERTIA OF AN AREA

The quantity to be discussed in this section is called *second moment of area* or *moment of inertia of an area*. It is used in the calculation of stresses in beams and columns and is often referred to simply as *moment of inertia*. It measures the effect of the cross-sectional shape of a beam on the beam's resistance to a bending moment. The usual units are in.[4].

Referring to Figure 9-1, we determine the moment of inertia of an area as follows:

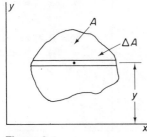

Figure 9-1

1. Divide the total area into incremental strips (ΔA) parallel to an axis, the x-axis in this case (Figure 9-1).

2. Multiply each area by the square of the distance to the x-axis.

$$I_x = y^2 \Delta A$$

3. Sum these terms for the total area.

$$I_x = \Sigma y^2 \Delta A \qquad (9\text{-}1)$$

33

Table 9-1 — Moments of Inertia and Radii of Gyration of Simple Areas

AREA	MOMENT OF INERTIA	RADIUS OF GYRATION
1. Rectangle 	$I_c = \dfrac{bh^3}{12}$ $I_x = \dfrac{bh^3}{3}$	$k_c = \dfrac{h}{\sqrt{12}}$ $k_x = \dfrac{h}{\sqrt{3}}$
2. Triangle 	$I_c = \dfrac{bh^3}{36}$ $I_x = \dfrac{bh^3}{12}$	$k_c = \dfrac{h}{\sqrt{18}}$ $k_x = \dfrac{h}{\sqrt{6}}$
3. Circle 	$I_c = \dfrac{\pi r^4}{4}$	$k_c = \dfrac{r}{2}$
4. Semicircle 	$I_c = 0.11\,r^4$ $I_x = \dfrac{\pi r^4}{8}$	$k_c = 0.264\,r$ $k_x = \dfrac{r}{2}$

To determine the moment of inertia about the y-axis, use vertical strips in the same manner; the total area is expressed as

$$I_y = \Sigma x^2 \Delta A \qquad (9\text{-}2)$$

For an area such as the one in Figure 9-1, the narrower the strip is, the greater the accuracy of the moment of inertia is. For this reason, the use of calculus generally yields the most accurate value.

Most areas considered will be simple geometric areas or composite areas of geometric shapes. Calculus has been used to determine moment of inertia formulae for these simple geometric areas (Table 9-1).

From Equations 9-1 and 9-2 where we had distance squared times area, it can be seen that the units for inertia of an area will be a unit of length to the 4th power, such as in.^{4} or mm^4.

The mm is a small unit of length and tends to give large numerical values. The use of meters is not necessarily that much better. For example, a value of $I = 625$ in.^{4} is equal to 0.00026 m^4 or 260,000,000 mm^4. Neither one of these terms is well suited for convenient calculation.

The cm is more convenient to use for inertia calculations. The final answer is converted from cm^4 to mm^4 or m^4.

$$1 \text{ cm} = 10 \text{ mm}$$
$$1 \text{ cm}^4 = (10 \text{ mm})^4$$
$$= 10^4 \text{ mm}^4$$
$$1 \text{ cm} = 1 \times 10^{-2} \text{ m}$$
$$1 \text{ cm}^4 = 1 \times 10^{-8} \text{ m}^4$$

9-2   PARALLEL AXIS THEOREM

As you may have noticed in Table 9-1, the moment of inertia was given about both the centroidal axis and the x-axis. We will now consider how the moment of inertia about the centroidal axis can be used to calculate the moment of inertia about any parallel axis, such as the x-axis.

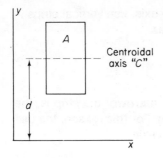

Figure 9-2

Suppose that in Figure 9-2 we wish to know the moment of inertia about the x-axis some distance d from the centroidal axis. Let

I_c = moment of inertia about the centroidal axis in mm⁴ or in.⁴

I_x = required moment of inertia about the x-axis in mm⁴ or in.⁴

d = perpendicular distance between parallel axes in mm or in.

A = total area in mm² or in.²

The *parallel axis theorem* states these terms as:

$$I_x = I_c + Ad^2 \qquad (9\text{-}3)$$

You may also see this equation referred to as the *transfer formula* since it transfers the moment of inertia from the centroidal axis to a parallel axis.

Example 9-1

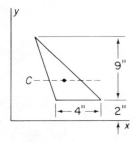

Figure 9-3

Determine the moment of inertia about the x-axis for the triangular area shown in Figure 9-3.

From Table 9-1, the moment of inertia about an axis through the centroid is:

$$I_c = \frac{bh^3}{36}$$

$$I_c = \frac{4(9)^3}{36}$$

$$I_c = 81 \text{ in.}^4$$

The distance from the base of the triangle to its centroid is

$$\frac{h}{3} = \frac{9}{3} = 3 \text{ in.} \qquad (\text{Figure 9-4})$$

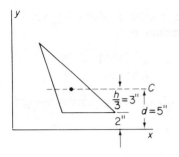

Figure 9-4

The distance d between parallel axes is 5 in.

Next, we find the area of the triangle.

$$A = \frac{1}{2}bh$$

$$= \frac{1}{2} \times 4 \times 9$$

$$A = 18 \text{ in.}^2$$

Now we apply the parallel axis equation (Equation 9-3).

$$I_x = I_c + Ad^2 \qquad (9\text{-}3)$$

$$= 81 + (18)(5)^2$$

$$= 81 + 450$$

$$I_x = 531 \text{ in.}^4$$

Example 9-2

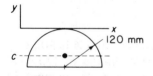

Figure 9-5

Determine the moment of inertia about the x-axis for the semicircular area shown in Figure 9-5.

From Table 9-1, the centroidal moment of inertia is

$$I_c = 0.11 r^4$$

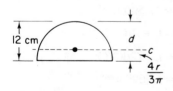

Figure 9-6

Using units of 12 cm rather than 120 mm (Figure 9-6), we get:

$$I_c = 0.11(12)^4$$

$$I_c = 2281 \text{ cm}^4$$

The distance from the base to the centroid is:

$$\frac{4r}{3\pi} = \frac{4 \times 12}{3 \times \pi} = 5.09 \text{ cm}$$

The centroidal moment of inertia will be transferred a distance of:

$$d = 12 - 5.09$$
$$= 6.91 \text{ cm}$$

Applying the parallel axis equation, we get:

$$I_x = I_c + Ad^2$$

$$= 2281 + \left(\frac{\pi \times 12^2}{2} \right)(6.91)^2$$

$$I_x = 13,100 \text{ cm}^4$$

or, since 1 cm^4 = 10^4 mm^4:

$$\underline{I_x = 131 \times 10^6 \text{ mm}^4}$$

To find the inertia about any new axis, one must always use the centroidal moment of inertia as the starting point. Moment of inertia cannot be transferred from any axis but the centroidal axis. The following example illustrates this point.

Example 9-3

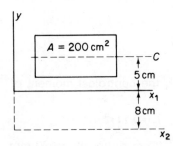

Figure 9-7

As shown in Figure 9-7, area A is 200 cm^2 and has a moment of inertia of 175×10^6 mm^4 about axis x_1. Find its moment of inertia about axis x_2.

We cannot go directly from x_1 to x_2 but must first find I_c and then, by means of the parallel axis equation, find I_{x_2}. Convert $I_{x_1} = 175 \times 10^6$ mm^4 = 17,500 cm^4.

$$I_{x_1} = I_c + Ad^2$$

$$17,500 = I_c + (200)(5)^2$$

$$I_c = 12,500 \text{ cm}^4$$

$$I_{x_2} = I_c + Ad^2$$

$$= 12,500 + (200)(13)^2$$

$$= 46,300 \text{ cm}^4$$

$$\underline{I_{x_2} = 463 \times 10^6 \text{ mm}^4}$$

9-3   MOMENT OF INERTIA OF COMPOSITE AREAS

The method of finding the moment of inertia of composite areas is very similar to the one used for centroids of composite areas. The usual sequence of steps is as follows:

1. Divide the composite area into simple areas.
2. For each simple area, find the moment of inertia about its centroidal axis.
3. Transfer each centroidal moment of inertia to a parallel reference axis.
4. The sum of the moments of inertia about the reference axis is the moment of inertia of the composite area. Any cut-out area has a negative moment; all other areas are positive.

Example 9-4

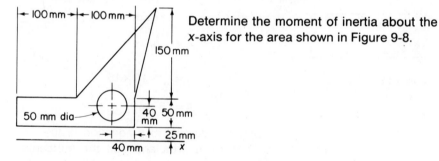

Determine the moment of inertia about the x-axis for the area shown in Figure 9-8.

Figure 9-8

Divide the total area into simple areas (Figures 9-9, 9-10, and 9-11). Find I_c and the area for each of the simple areas.

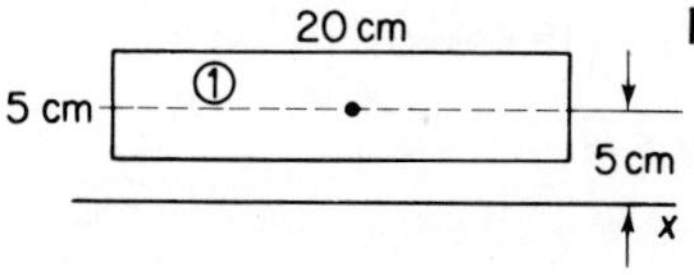

Figure 9-9

Rectangle (Figure 9-9):

$$I_c = \frac{bh^3}{12}$$

$$= \frac{(20)(5)^3}{12}$$

$$I_c = 208.3 \text{ cm}^4$$

$$A = 100 \text{ cm}^2$$

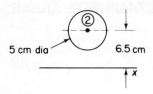

5 cm dia   **6.5 cm**

x

Figure 9-10

Circle (Figure 9-10):

$$I_c = \frac{\pi r^4}{4}$$

$$= \frac{\pi (2.5)^4}{4}$$

$$I_c = 30.7 \text{ cm}^4$$

$$A = \frac{\pi d^2}{4}$$

$$A = 19.6 \text{ cm}^2$$

Triangle (Figure 9-11):

15 cm

③ $\frac{h}{3} = 5$ cm

←—10 cm—→

7.5 cm

x

Figure 9-11

$$I_c = \frac{bh^3}{36}$$

$$= \frac{(10)(15)^3}{36}$$

$$I_c = 937 \text{ cm}^4$$

$$A = \frac{1}{2} bh$$

$$= \frac{1}{2} \times 10 \times 15$$

$$A = 75 \text{ cm}^2$$

Apply the parallel axis equation (Equation 9-3) to each area and add the moments about the x-axis. Note that I_x of the circle is negative since it is a cut-out area.

$$I_x = I_{x_1} - I_{x_2} + I_{x_3}$$

$$= (I_c + Ad^2)_1 - (I_c + Ad^2)_2 + (I_c + Ad^2)_3$$

$$= \left[208.3 + (100)(5)^2 \right]$$

$$- \left[30.7 + (19.6)(6.5)^2 \right]$$

$$+ \left[937 + (75)(12.5)^2 \right]$$

$$= 2708 - 859 + 12{,}650$$

$$= 14{,}500 \text{ cm}^4$$

$$I_x = 145 \times 10^6 \text{ mm}^4$$

Example 9-5

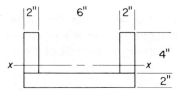

Figure 9-12

Determine the moment of inertia about the horizontal centroidal axis of the composite area in Figure 9-12.

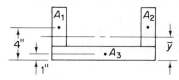

Figure 9-13

The solution is similar to that of Example 9-4; here, however, we must first find the location of the centroidal axis x-x (Figure 9-13).

$$A = A_1 + A_2 + A_3$$
$$= 8 + 8 + 20$$
$$A = 36 \text{ in.}^2$$
$$A\bar{y} = A_1 y_1 + A_2 y_2 + A_3 y_3$$
$$36\bar{y} = (8 \times 4) + (8 \times 4) + (20 \times 1)$$
$$\bar{y} = \frac{84}{36} = 2.33 \text{ in.}$$

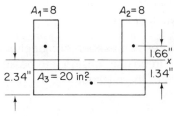

Figure 9-14

Determine the centroidal inertia for each area and transfer it to the x-axis (Figure 9-14). This can be done with the one equation.

$$I_x = (I + Ad^2)_1 + (I + Ad^2)_2 + (I + Ad^2)_3$$
$$= \frac{2(4)^3}{12} + 8(1.66)^2 + \frac{2(4)^3}{12} + 8(1.66)^2$$
$$+ \frac{10(2)^3}{12} + 20(1.34)^2$$
$$= 10.67 + 22.05 + 10.67 + 22.05$$
$$+ 6.67 + 35.9$$
$$I_x = 108 \text{ in.}^4$$

9-4 RADIUS OF GYRATION

The *radius of gyration, k*—the distance from the centroidal axis of an area at which the entire area could be concentrated and still have the same moment of inertia—is defined by the following sequence of steps:

1. Calculate I_c for an area A.

2. Assume that area A is concentrated at some distance k from the centroidal axis in such a manner that we still have a moment of inertia equal to I_c.

3. Recall that area moment of inertia was defined by the equation, moment of inertia = (area) (distance)^2. Therefore, we have:

$$I_c = Ad^2$$

or

$$k = \sqrt{\frac{I_c}{A}}$$

where

$$I_c = \text{moment of inertia in in.}^4, \text{mm}^4, \text{or cm}^4$$
$$A = \text{total area in in.}^2, \text{mm}^2, \text{or cm}^2$$
$$k = \text{radius of gyration in in. mm, or cm}$$

The radius of gyration is a mathematical expression used in conjunction with the slenderness ratio in column design. In column design, the beam will fail by bending about the centroidal axis of the minimum k.

Example 9-6

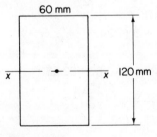

Figure 9-15

Determine the radius of gyration about the x-axis of the area shown in Figure 9-15.

From Table 9-1, the equation for centroidal moment inertia is:

$$I_c = \frac{bh^3}{12}$$

Using cm rather than mm, we get:

$$I_c = \frac{6 \times (12)^3}{12} = 864 \text{ cm}^4$$

$$A = 6 \times 12 = 72 \text{ cm}^2$$

The radius of gyration is the distance from the x-axis at which the area of 72 cm² can be concentrated and still have a moment of inertia of 864 cm⁴ about the x-axis.

$$k = \sqrt{\frac{I_c}{A}}$$

$$= \sqrt{\frac{864}{72}}$$

$$= 3.46 \text{ cm}$$

$$\underline{k = 34.6 \text{ mm}}$$

Example 9-7

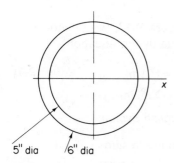

Figure 9-16

5" dia   6" dia

A tube is used as a column; it has the cross-sectional area shown in Figure 9-16. Determine the radius of gyration about the x-axis.

Find I_x of the composite area by considering the area due to the 5-ft diameter to be negative.

Using the inertia formula from Table 9-1, we have:

$$I_x = \frac{\pi r_1^4}{4} - \frac{\pi r_2^4}{4}$$

$$= \frac{\pi}{4}(3^4 - 2.5^4)$$

$$I_x = 33 \text{ in.}^4$$

Calculate the cross-sectional area.

$$A = \frac{\pi d_1^2}{4} - \frac{\pi d_2^2}{4}$$

$$= \frac{\pi}{4}(6^2 - 5^2)$$

$$A = 8.63 \text{ in.}^2$$

$$k = \sqrt{\frac{I}{A}}$$

$$= \sqrt{\frac{33}{8.63}}$$

$$\underline{k = 1.95 \text{ in.}}$$

9-5   MASS MOMENT OF INERTIA

Mass moment of inertia is a measure of resistance to rotational acceleration and will be used later in dynamics of rotational motion.

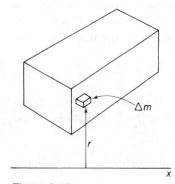

Figure 9-17

Referring to Figure 9-17, we define mass moment of inertia by the equation:

$$\Delta I = r^2 \Delta m \qquad (9\text{-}4)$$

$$I = \Sigma r^2 \Delta m$$

where in the English System

$$m = \text{mass in slugs}$$

$$= \frac{W}{g} = \frac{\text{weight}}{\text{gravity acceleration}}$$

$$= \frac{\text{lb}}{\text{ft/sec}^2} = \frac{\text{lb-sec}^2}{\text{ft}}$$

r = perpendicular distance from the axis to Δm in ft

I_x = mass moment of inertia in ft-lb-sec^2 or slug-ft^2

All dimensions used in connection with mass moment of inertia in the English System must be in feet since mass has units of slugs.

In the SI system, the units for Equation 9-4 are:

$$m = \text{mass in kg}$$
$$r = \text{distance in m}$$
$$I_x = \text{mass moment of inertia in kg} \cdot \text{m}^2$$

Recall that

$$\text{mass in kg} = \frac{\text{force of gravity in N}}{\text{acceleration of gravity of 9.81 m/s}^2} \, .$$

While all final answers should be shown in units of kg·m², to avoid cumbersome numbers, all intermediate calculations can use cm or mm with the conversion to m being made for the final answer.

Mass moment of inertia values can be converted from English to SI units by using the conversion factor 1 ft-lb-sec² = 1.356 kg·m².

As with the other types of inertia of areas that we have discussed, mass moment of inertia is accurately computed by means of calculus. Examples of calculated mass inertia of simple shapes are listed in Table 9-2. We will be considering composite bodies that can be broken into these shapes.

Example 9-8

Determine the mass moment of inertia about the longitudinal axis of a shaft 80 mm in diameter that has a mass of 25 kg.

From Table 9-2:

$$I = \frac{1}{2} mr^2$$

where, using meters, we get:

$$m = 25 \text{ kg}$$

$$r = 40 \text{ mm} = 0.04 \text{m}$$

$$I = \frac{1}{2} \times 25 \times (4 \times 10^{-2})^2$$

$$= \frac{25}{2} \times 16 \times 10^{-4}$$

$$= 200 \times 10^{-4}$$

$$\underline{I = 0.02 \text{ kg} \cdot \text{m}^2}$$

or using mm, we get:

$$m = 25 \text{ kg}$$
$$r = 40 \text{ mm}$$
$$I = \frac{1}{2} \times 25 \times (40)^2$$
$$I = 20,000 \text{ kg} \cdot \text{mm}^2$$

but $1 \text{ m}^2 = 10^6 \text{ mm}^2$. Therefore:

$$\underline{I = 0.02 \text{ kg} \cdot \text{m}^2}$$

Table 9-2

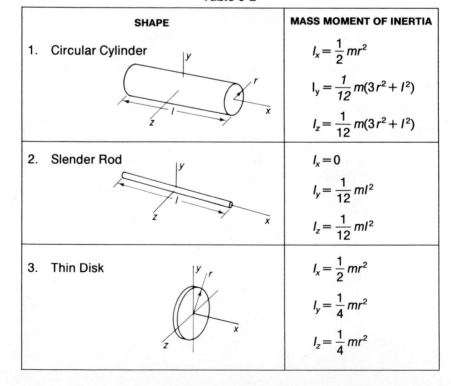

SHAPE	MASS MOMENT OF INERTIA
1. Circular Cylinder	$I_x = \frac{1}{2} mr^2$ $I_y = \frac{1}{12} m(3r^2 + l^2)$ $I_z = \frac{1}{12} m(3r^2 + l^2)$
2. Slender Rod	$I_x = 0$ $I_y = \frac{1}{12} ml^2$ $I_z = \frac{1}{12} ml^2$
3. Thin Disk	$I_x = \frac{1}{2} mr^2$ $I_y = \frac{1}{4} mr^2$ $I_z = \frac{1}{4} mr^2$

Table 9-2 (continued)

SHAPE	MASS MOMENT OF INERTIA
4. **Right Circular Cone**	$I_x = \dfrac{3}{10} mr^2$ $I_y = \dfrac{3}{5} m(\dfrac{1}{4} r^2 + h^2)$ $I_z = \dfrac{3}{5} m(\dfrac{1}{4} r^2 + h^2)$
5. **Sphere**	$I_x = \dfrac{2}{5} mr^2$ $I_y = \dfrac{2}{5} mr^2$ $I_z = \dfrac{2}{5} mr^2$
6. **Hemisphere**	$I_x = \dfrac{2}{5} mr^2$ $I_y = \dfrac{2}{5} mr^2$ $I_z = \dfrac{2}{5} mr^2$
7. **Rectangular Prism**	$I_x = \dfrac{1}{12} m(a^2 + b^2)$ $I_y = \dfrac{1}{12} m(a^2 + l^2)$ $I_z = \dfrac{1}{12} m(b^2 + l^2)$
8. **Hollow Cylinder**	$I_x = \dfrac{1}{2} m(r_1^2 + r_2^2)$

9-6 MASS MOMENT OF INERTIA OF COMPOSITE BODIES

Like moment of inertia of area, mass moment of inertia can be transferred from the centroidal axis to any other axis of rotation by means of the Equation 9-5 below.

$$I = I_c + md^2 \tag{9-5}$$

where

I = mass moment of inertia about some new axis in ft-lb-sec² or kg·m²

I_c = mass moment of inertia about the centroidal axis in ft-lb-sec² or kg·m²

m = mass in slugs or kg

d = perpendicular distance between axes in ft. or m

Example 9-9

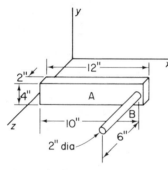

Figure 9-18

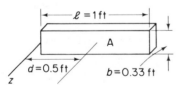

Figure 9-19

In the composite body shown in Figure 9-18, A weighs 25 lb and B weighs 10 lb. Calculate the mass moment of inertia about the z-axis.

Divide the composite body into bodies A and B and calculate moment of inertia of each about the x-axis, i.e., I_{z_A} and I_{z_B}.

Applying the appropriate equation from Table 9-2 to our condition of body A (Figure 9-19), we obtain:

$$I_c = \frac{1}{12} m(b^2 + l^2)$$

$$= \frac{1}{12} \times \frac{W}{g}(b^2 + l^2)$$

$$= \frac{1}{12} \times \frac{25}{32.2}(0.33^2 + 1^2)$$

$$I_c = 0.072 \text{ ft-lb-sec}^2$$

$$I_{z_A} = I_c + md^2$$

$$= 0.072 + \frac{25}{32.2} \times (0.5)^2$$

$$I_{z_A} = 0.266 \text{ ft-lb-sec}^2$$

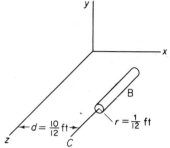

Figure 9-20

Applying the same procedure to body B (Figure 9-20), we obtain:

$$I_c = \frac{1}{2} mr^2$$

$$= \frac{1}{2} \times \frac{10}{32.2} \times \left(\frac{1}{12}\right)^2$$

$$I_c = 0.0011 \text{ ft-lb-sec}^2$$

$$I_{z_B} = I_c + md^2$$

$$= 0.0011 + \frac{10}{32.2} \times \left(\frac{10}{12}\right)^2$$

$$I_{z_B} = 0.216 \text{ ft-lb-sec}^2$$

The total moment of inertia with respect to the z-axis is

$$I_z = I_{z_A} + I_{z_B}$$
$$= 0.266 + 0.216$$
$$I_z = 0.482 \text{ ft-lb-sec}^2$$

Example 9-10

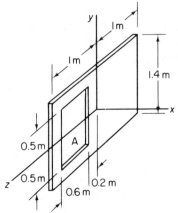

Figure 9-21

Calculate the mass moment of inertia about the x-axis for the thin rectangular plate shown in Figure 9-21. The plate had a mass of 5 kg before the hole material (1.07 kg) was cut out.

The method required here is treatment of the hole as a negative mass moment of

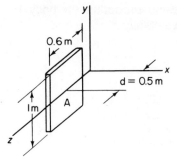

Figure 9-22

inertia. Figure 9-22 illustrates the hole treated as an equivalent piece of plate.

$$I_c = \frac{1}{12}m(a^2 + b^2)$$

$$= \frac{1}{12} \times 1.07[(0.6)^2 + (1)^2]$$

$$I_c = 0.121 \text{ kg} \cdot \text{m}^2$$

$$I_{x_A} = I_c + md^2$$

$$= 0.121 + 1.07(0.5)^2$$

$$I_{x_A} = 0.388 \text{ kg} \cdot \text{m}^2$$

Find I_x for the plate without the hole.

$$I_x = I_c = \frac{1}{12}m(a^2 + b^2)$$

$$= \frac{1}{12} \times 5[(2)^2 + (1.4)^2]$$

$$I_x = 2.48 \text{ kg} \cdot \text{m}^2$$

The moment of inertia of the plate with the hole is:

$$I_x = 2.48 - 0.388$$

$$\underline{I_x = 2.1 \text{ kg} \cdot \text{m}^2}$$

9-7   RADIUS OF GYRATION OF BODIES

Once again—just as in the case of the radius of gyration of areas—the radius of gyration is defined as:

$$k = \sqrt{\frac{I}{m}} \qquad (9\text{-}6)$$

where

$m =$ mass in slugs or kg

$k =$ radius of gyration in ft or m

$I =$ mass moment of inertia in ft-lb-sec^2 or kg$\cdot$m^2

For the radius of gyration of a simple body, I_c is used to find the moment of inertia. To obtain the radius of gyration of a composite body, substitute the final I value into Equation 9-6. You cannot add the individual radii of gyration of the simple bodies to get the radius of gyration of a composite body.

Another often used form of Equation 9-6 is:

$$I = k^2 m \qquad \text{(9-7)}$$

PROBLEMS

9-1–9-6 Determine the moment of inertia about the x-axis of each of the areas shown in Figures P9-1 to P9-6.

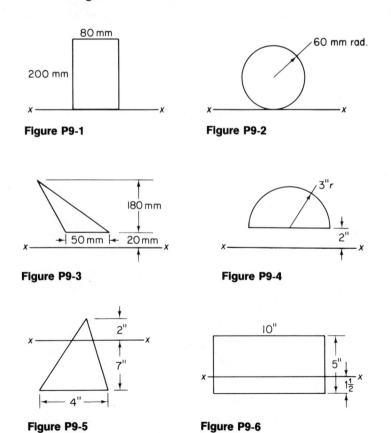

Figure P9-1

Figure P9-2

Figure P9-3

Figure P9-4

Figure P9-5

Figure P9-6

9-7 Determine the moment of inertia about the *y*-axis of the area shown in Figure P9-7.

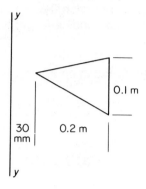

Figure P9-7

9-8 The moment of inertia $I_{x_1} = 10.5$ in.^{4} for the area shown in Figure 9-8. Find I_{x_2}.

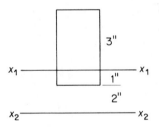

Figure P9-8

9-9 Determine the moment of inertia I_x for the area shown in Figure P9-9.

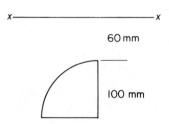

Figure P9-9

9-10 The moment of inertia $I_{x_1} = 1203$ in.^{4} for the area shown in Figure P9-10. Find I_{x_2}.

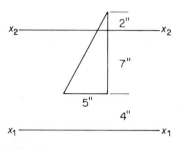

Figure P9-10

9-11–9-14 Determine the moments of inertia about the *x*-axis for the areas shown in Figures P9-11 to P9-14.

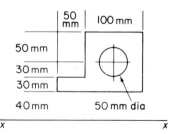

Figure P9-11

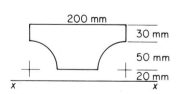

Figure P9-12

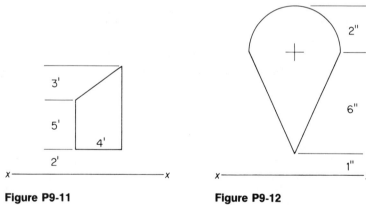

Figure P9-13

Figure P9-14

9-15   Determine the moment of inertia of the area shown in Figure P9-15 about its centroidal *x*-axis.

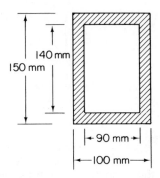

Figure P9-15

9-16   Determine the radius of gyration for Problem 9-15.

9-17   The plate shown in Figure P9-17 has a slot cut in it. Determine the area moment of inertia about its centroidal *x*-axis.

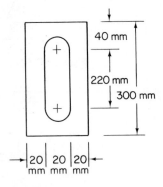

Figure P9-17

9-18   Determine the radius of gyration for Problem 9-17.

9-19–9-28 Determine the moment of inertia about the centroidal *x*-axis of the areas shown in Figures P9-19 to P9-28.

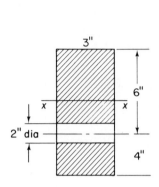

Figure P9-19

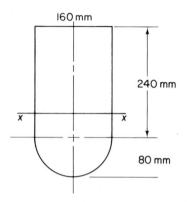

Figure P9-20

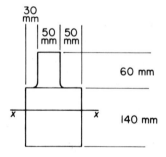

Figure P9-21

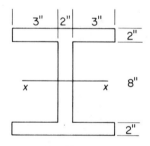

Figure P9-22

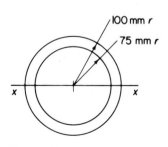

Figure P9-23

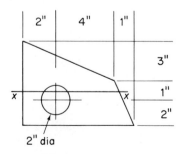

Figure P9-24

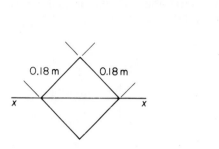

Figure P9-25

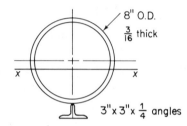

Figure P9-26

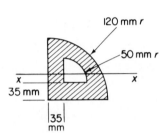

Figure P9-27

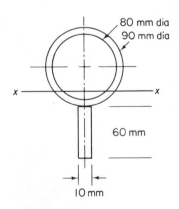

Figure P9-28

9-29 Determine the moment of inertia of the area shown in Figure 9-29 about its centroidal *x*-axis.

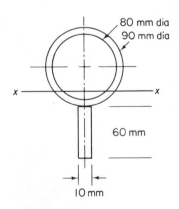

80 mm dia
90 mm dia

60 mm

10 mm

Figure P9-29

9-30 Determine the radius of gyration for Problem 9-29.

9-31 Determine the moment of inertia about the centroidal x-axis of the fabricated area shown in Figure P9-31.

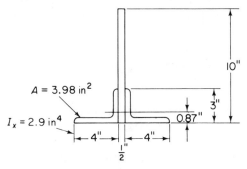

Figure P9-31

9-32 Determine the moment of inertia about the centroidal x-axis of the cross-sectional area of the fabricated beam shown in Figure P9-32.

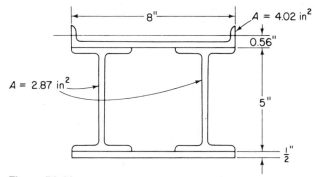

Figure P9-32

9-33 The cast frame of a press is subjected to a bending stress and has a cross-sectional area as shown in Figure P9-33. As a first step in the calculation of bending stress, determine the moment of inertia about the centroidal y-axis.

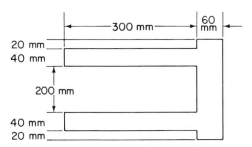

Figure P9-33

9-34 Determine the radius of gyration of the area shown in Figure P9-20 with respect to the x-axis.

9-35 Determine the radius of gyration of the area shown in Figure P9-21 with respect to the x-axis.

9-36 Determine the radius of gyration of the area shown in Figure P9-22 with respect to the x-axis.

9-37 Determine the radius of gyration of the area shown in Figure P9-26 with respect to the x-axis.

9-38 Determine the mass moment of inertia and the radius of gyration about the centroidal axis of a sphere 2 ft in diameter and weighing 64.4 lb.

9-39 Determine the mass moment of inertia and the radius of gyration about the centroidal longitudinal axis of a shaft that has a mass of 100 kg and a diameter of 120 mm.

9-40 Determine the mass moment of inertia of a solid cylinder, 4 ft in diameter, about an axis on the surface of the cylinder and parallel to its centroidal axial axis. The cylinder weighs 96.6 lb.

9-41 A slender rod 0.6 m long rotates about an axis perpendicular to its length and 0.14 m from its center of gravity. If the rod has a mass of 8 kg, determine its mass moment of inertia about this axis.

9-42 The right circular cone in Figure P9-42 has a mass of 90 kg. Determine the mass moment of inertia about the x-axis.

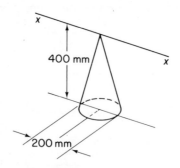

Figure P9-42

9-43   A flywheel can be considered as being composed of a thin disk and a rim. The rim weighs 322 lb and has diameters of 24 in. and 30 in. The disk weighs 64.4 lb. Determine the mass moment of inertia about the centroidal axis about which the flywheel rotates.

9-44   In the system shown in Figure P9-44, the sphere weighs 10 lb, the rod weighs 8 lb, and the plate weighs 4 lb. Determine the mass moment of inertia about the y-axis.

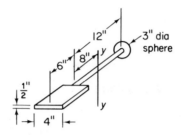

Figure P9-44

9-45   The plate in Figure P9-45 has a mass of 3000 kg/m³. Determine the mass moment of inertia about the x-axis.

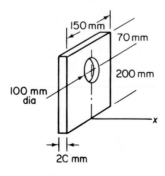

Figure P9-45

9-46 The shape in Figure P9-46 weighs 0.2 lb/in.³. Determine the radius of gyration about the y-axis. The volume of a right circular cone is $\frac{1}{3}\pi r^2 h$.

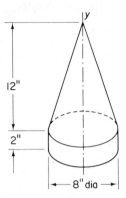

Figure P9-46

9-47 The shaft of a shredder has cutter blades welded to the shaft as shown in Figure P9-47. The blades are 10 mm thick. The shaft and blades have a mass of 8000 kg/m³. Determine the mass moment of inertia about the longitudinal centroidal axis.

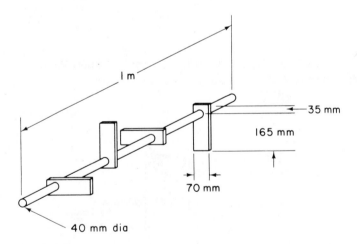

Figure P9-47

DYNAMICS

10

KINEMATICS−
RECTILINEAR MOTION

10-1 INTRODUCTION

Previously, we were concerned with *statics*—bodies at rest or with uniform velocity. We now come to *dynamics*—the study of bodies in motion. Dynamics consists of kinematics and kinetics.

Kinematics is the analysis of the geometry of motion without concern for the forces causing the motion; it involves quantities such as displacement, velocity, acceleration, and time.

Kinetics is the study of motion and the forces associated with motion; it involves the determination of the motion resulting from given forces.

In our study of kinematics, we will consider motion in one plane only; such motion will be one of three types.

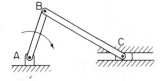

Figure 10-1

1. *Rectilinear* or *translational motion*: the particle or body moves in a straight line and does not rotate about its center of mass. The piston, pin C in Figure 10-1, has rectilinear or straight line motion.

2. *Circular motion*: a particle follows the path of a perfect circle. In Figure 10-1, pin B on the end of arm AB has circular motion.

3. *General plane motion*: a particle may follow a path that is neither straight nor circular. This also applies to a body that may have both rotating and rectilinear motion simultaneously.

Figure 10-2

Link BC has both rectilinear and circular motion since it partially rotates as it moves from left to right. General plane motion may be even more random and undefined, as shown by the particle moving from A to B in Figure 10-2.

Only particles will be considered in the kinematics of this chapter. A *particle* here refers either to a small concentrated object or to a large object whose center of mass has motion identical to that of all of the other parts of the object.

A half-loaded tanker truck could not be considered to possess particle motion since the center of mass changes due to the tank's moving on its springs and the load's shifting inside the tank. When an object has rotation about its center of mass, it too cannot possess particle motion since not all portions of the object have identical motion. This type of motion will be considered in Chapter 12.

The term ''particle'' does not, therefore, necessarily mean a very small object; it could also refer to a very large object if all portions of that object have the same motion as its center of mass.

10-2 DISPLACEMENT

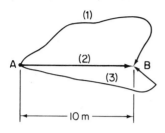

Figure 10-3

There is a distinct difference between *distance* and *displacement*. Distance is a scalar quantity, and displacement is a vector quantity. To travel from point A to B (Figure 10-3), any one of three paths may be used. Each path involves a different distance, but they all have the same displacement, 10 m to the right. Displacement is merely the difference between original position and some later position.

Equations that involve velocity, acceleration, and time will also have displacement values.

Example 10-1

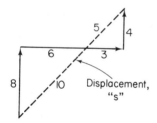

Figure 10-4

A car is driven 8 km north, 9 km east, and then another 4 km north. Calculate the displacement of the car and the distance travelled.

The total distance is, naturally, $8+9+4=21$ km. The path travelled can be shown as in Figure 10-4. By the geometry of similar triangles, the lengths of the remaining sides of the two triangles can be found. The displacement of the car is 15 km $\angle_4^4$.

10-3 VELOCITY

There is also a basic distinction between *speed* and *velocity*. Speed is a scalar quantity; velocity is a vector quantity.

Speed is the change of distance per unit of time, such as meters per second (m/s), kilometers per hour (km/h), feet per minute (ft/min), or feet per second (ft/sec). None of these units of speed indicates any direction.

Velocity, however, indicates both speed and direction. Velocity is the rate of change of displacement with respect to time and has the same units as speed, i.e., m/s, km/h, ft/sec, and ft/min. In equation form, we have:

$$v = \frac{s}{t} = \frac{\Delta s}{\Delta t} \tag{10-1}$$

where

$$v = \text{velocity (average)}$$
$$s = \text{displacement}$$
$$t = \text{time}$$
$$\Delta s = s_2 - s_1$$
$$\Delta t = t_2 - t_1$$

The velocity obtained by this equation is an average velocity, since there are no values given for intermediate values of displacement and time. A driver of a car that travels 60 km from point A to B in one hour may have travelled at a constant speed of 60 km/h or may have stopped for a half-hour and then travelled at 120 km/h. The velocity equation tells us that his end result is an *average velocity* of 60 km/h.

Example 10-2

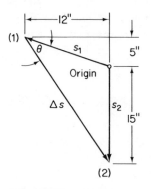

Figure 10-5

The path followed by a pin in a printing press mechanism is shown in Figure 10-5. Starting from the origin, it reaches point (1) in 2 seconds and then point (2) after a total elapsed time of 3 seconds. Determine the average velocity from point (1) to (2).

Use Equation 10-1 to solve.

$$v = \frac{\Delta s}{\Delta t} = \frac{s_2 - s_1}{t_2 - t_1}$$

From Figure 10-5:

$$\Delta s = \sqrt{(12)^2 + (20)^2}$$
$$= 23.3 \text{ in.}$$

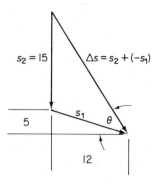

Figure 10-6

Or, velocity may also be solved for by means of the vector triangle in Figure 10-6 ($s_2 - s_1 = 23.3$ in.).

$$v = \frac{23.3}{3-2}$$

$$v = 23.3 \text{ in./sec } \sqrt{59°}$$

Since

$$\tan \theta = \frac{20}{12} = 1.667$$

$$\theta = 59°$$

(Remember that this value represents only an average velocity.)

10-4 ACCELERATION

Acceleration is the rate of change of velocity with respect to time. As was discussed previously, velocity has both direction and magnitude. A change of velocity either in direction or in magnitude constitutes acceleration. The velocity change with which we are most familiar is that of magnitude. A car whose speed increases from 30 to 60 km/h is said to accelerate. If the car now changes direction while maintaining a speed of 60 km/h, it again experiences acceleration; this time, acceleration is due to the change in direction. Both types of acceleration can be calculated by means of the formula:

$$a = \frac{\Delta v}{\Delta t} \qquad (10\text{-}2)$$

where

$a =$ acceleration (average)

$\Delta v =$ velocity change

$\quad = v_2 - v_1$

$\Delta t =$ time for velocity change

$\quad = t_2 - t_1$

Substituting the units of velocity and time into this equation, we see that acceleration is expressed in terms of displacement/time/time. The common units are mm/s², m/s², in./sec², and ft/sec². Acceleration is a vector quantity that has the same direction as the velocity change. Equation 10-2 indicates an average acceleration for a given time period.

Example 10-3

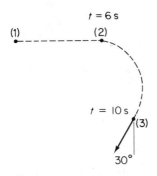

Figure 10-7

The car in Figure 10-7 starts from rest at point (1) and uniformly accelerates due east for 6 s, reaching a speed of 40 km/h at point (2). Maintaining the speed of 40 km/h, it reaches point (3) at $t=10$ s, travelling a direction of south 30° west. Determine the acceleration (a) from point (1) to point (2), and (b) from point (2) to point (3).

The first important step is to obtain uniform units: convert km/h to m/s.

$$40 \text{ km/h} = \frac{40 \times 1000}{60 \times 60} = 11.1 \text{ m/s}$$

Applying the acceleration equation (Equation 10-2) between points (1) and (2), we get:

$$a = \frac{\Delta v}{\Delta t} = \frac{v_2 - v_1}{t_2 - t_1}$$

$$= \frac{11.1 - 0}{6 - 0}$$

$$a = 1.85 \text{ m/s}^2$$

Between points (2) and (3):

$$a = \frac{\Delta v}{\Delta t} = \frac{v_3 - v_2}{t_3 - t_2}$$

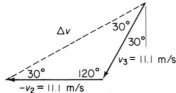

Figure 10-8

To find $v_3 - v_2$ vectorially, we add a negative v_2 to v_3; that is, $v_3 + (-v_2)$ (Figure 10-8).

By using the sine law, we get:

$$\frac{\Delta v}{\sin 120°} = \frac{11.1}{\sin 30°}$$

$$\Delta v = \frac{11.1 \times 0.866}{0.5}$$

$$\Delta v = 19.2 \text{ m/s} \quad {}^{30} \downarrow$$

Substituting this value into the acceleration equation, we get:

$$a = \frac{v_3 - v_2}{t_3 - t_2}$$

$$= \frac{19.2}{10 - 6}$$

$$\underline{a = 4.8 \text{ m/s}^2 \quad {}^{\overline{30\ \circ\ \downarrow}}}$$

Note here that the direction of the acceleration vector must be the same as the direction of the velocity change, Δv. Remember that these are average velocities and average accelerations. The instantaneous values of acceleration between points (2) and (3) are constantly changing direction.

10-5   RECTILINEAR MOTION WITH UNIFORM ACCELERATION

Motion in a straight line with uniform or constant acceleration is relatively common and easy to analyze. A free-falling body is one example of this type of motion since the acceleration due to gravity (g) is assumed constant, i.e., at $g = 9.81$ m/s^2 (32.2 ft/sec^2) at sea level.

The three equations that relate the variables of time, displacement, velocity, and acceleration to each other are:

$$s = v_o t + \frac{1}{2} a t^2 \qquad\qquad (10\text{-}3)$$

$$v = v_o + at \qquad\qquad (10\text{-}4)$$

$$v^2 = v_o^2 + 2as \qquad\qquad (10\text{-}5)$$

where, in common units:

$s =$ displacement in m or ft

$v_o =$ initial velocity in m/s or ft/sec

$v =$ final velocity in m/s or ft/sec

$a =$ constant acceleration in m/s^2 or ft/sec^2

$t =$ time in seconds

Keep the following points in mind when you use these equations:

1. Acceleration, although it may be in any direction, must be constant. Constant velocity is a special case in which acceleration is constant at zero.
2. A free-falling body has an acceleration of $a = g = 9.81$ m/s² (32.2 ft/sec²).
3. An object that decelerates or slows down is treated as having a negative acceleration.
4. Designate the direction that is to be positive—the direction of the initial velocity or displacement is often used as the positive direction.

Example 10-4

An object dropped from the top of a building strikes the ground 7 s later. What was the height of the building and with what velocity did the object strike the ground?

A good rule to follow with this type of problem is to tabulate all given information as follows since the equation that should be used may become more obvious.

$$t = 7 \text{ s}$$
$$a = 9.81 \text{ m/s}^2$$
$$s = ?$$
$$v = ?$$
$$v_o = 0$$

Using Equation 10-3, we obtain:

$$s = v_o t + \frac{1}{2} a t^2$$
$$= 0 + \frac{1}{2} \times 9.81 \times (7)^2$$
$$\underline{s = 240 \text{ m}}$$

Using Equation 10-4, we obtain:

$$v = v_o + at$$
$$= 0 + 9.81 \times 7$$
$$\underline{v = 68.7 \text{ m/s} \downarrow}$$

Example 10-5

A helicopter accelerates uniformly upward at 1 m/s² to a height of 300 m. By the time it reaches 350 m, it has decelerated to zero vertical velocity. It then accelerates horizontally at 4 m/s² to a velocity of 15 m/s. Determine the total time required for this sequence.

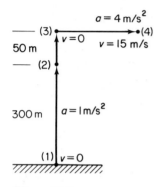

Figure 10-9

A suggestion for the solution here is to draw a sketch of the several stages of flight, as in Figure 10-9, and to label the given information.

Now calculate the time for each stage. Stage (1) to (2):

$$s = 300 \text{ m}$$
$$v_o = 0$$
$$a = 1 \text{ m/s}^2$$
$$s = v_o t + \frac{1}{2} a t^2$$
$$300 = 0 + \frac{1}{2} \times 1 \times t^2$$
$$t = 24.5 \text{ s}$$

Let the velocity at (2) be v_2.

$$v = v_o + at$$
$$v_2 = 0 + 1 \times 24.5$$
$$v_2 = 24.5 \text{ m/s} \uparrow$$

Stage (2) to (3):

$$v_o = 24.5 \text{ m/s} \uparrow$$
$$s = 50 \text{ m}$$
$$v = 0$$
$$v_2 = v_o^2 + 2as$$
$$0 = (24.5)^2 + 2 \times a \times 50$$
$$a = -6 \text{ m/s}^2 \uparrow$$

(The minus sign indicates a deceleration in

the direction of the arrow.)

$$v = v_o + at$$
$$0 = 24.5 - 6 \times t$$
$$\underline{t = 4.08 \text{ s}}$$

Stage (3) to (4):

$$v_o = 0$$
$$v = 15 \text{ m/s} \rightarrow$$
$$a = 4 \text{ m/s}^2 \rightarrow$$
$$v = v_o + at$$
$$15 = 0 + 4t$$
$$\underline{t = 3.75 \text{ s}}$$

total time elapsed $= 24.5 + 4.08 + 3.75$

total time elapsed $= 32.3 \text{ s}$

Example 10-6

A particle starting from rest and travelling to the right in a straight line accelerates uniformly to a velocity of 30 ft/sec in 3 sec. It then uniformly decelerates so that, from its initial place of rest to its final position, the displacement is 15 ft to the right and the total distance travelled is 155 ft. Determine the total time interval and the final velocity.

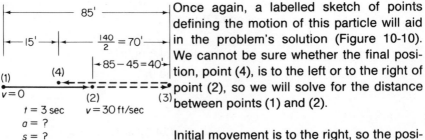

Figure 10-10

Once again, a labelled sketch of points defining the motion of this particle will aid in the problem's solution (Figure 10-10). We cannot be sure whether the final position, point (4), is to the left or to the right of point (2), so we will solve for the distance between points (1) and (2).

Initial movement is to the right, so the positive direction throughout the solution will be to the right.

Between (1) and (2):

$$v_o = 0$$
$$v = 30 \, \text{ft/sec}$$
$$t = 3 \, \text{sec}$$
$$a = ?$$
$$s = ?$$
$$v = v_o + at$$
$$30 = 0 + a \times 3$$
$$a = 10 \, \text{ft/sec}^2 \rightarrow$$
$$s = v_o t + \frac{1}{2} at^2$$
$$= 0 + \frac{1}{2} \times 10 \times (3)^2$$
$$s = 45 \, \text{ft}$$

To get from point (2) to the final position, point (4), the particle must reverse direction; that is, it must decelerate. Since at point (2) it is travelling 30 ft/sec to the right, it must decelerate to zero velocity at some point—point (3)—and, at this same deceleration, return to the left, arriving at point (4) with some final velocity.

If the total distance travelled from points (1) to (4) is 155 ft, then the distance between (3) and (4) is $155 - 15/2 = 70$ ft (Figure 10-10). The distance between (2) and (3) is then $(15 + 70) - (45) = 40$ ft.

Solve for the deceleration rate between points (2) and (3).

$$v_o = 30 \, \text{ft/sec}$$
$$v = 0$$
$$a = ?$$
$$s = 40 \, \text{ft}$$
$$v^2 = v_o^2 + 2as$$
$$0 = (30)^2 + 2 \times a \times 40$$
$$a = -11.25 \, \text{ft/sec}^2$$

You could now solve for individual velocities and times between points (2) and (3) and points (3) and (4), but a shorter method would be to calculate directly the velocity between points (2) and (4).

$$s = -30\,\text{ft}$$
$$v_o = 30\,\text{ft/sec}$$
$$v = ?$$
$$a = -11.25\,\text{ft/sec}^2$$
$$v^2 = v_o^2 + 2as$$
$$v^2 = (30)^2 + (2)(-11.25)(-30)$$
$$\underline{v = 39.7\,\text{ft/sec} \leftarrow}$$

Since *v* was squared, we cannot know the sign of this answer, but by observation we know that the direction of the velocity is to the left. Once again, between points (2) and (4):

$$v = v_o + at$$
$$-39.7 = 30 - 11.25t$$
$$t = 6.2\,\text{sec}$$
$$\text{total time} = 3 + 6.2$$
$$\underline{\text{total time} = 9.2\,\text{sec}}$$

10-6   PROJECTILES

A golf ball driven down the fairway, a baseball hit for a home run, and a rocket fired from a launching pad are all examples of a projectile tracing a path (*trajectory*) during its travel. This *projectile motion* consists of two rectilinear motions occurring simultaneously; projectile motion is comprised of both vertical and horizontal motion. Each of these motions can be represented by displacement, velocity, and acceleration vectors; each motion can therefore be treated separately, or they can be added together vectorially.

The main assumption here will be that of zero air resistance. The only factors affecting projectile motion will thus be initial velocity, the projectile's direction, and the pull of gravity. Acceleration is constant in both directions—zero horizontally and approximately 9.81 m/s² (32.2 ft/sec²) vertically.

The direction of displacements, velocities, and accelerations will be shown by the previously used sign convention, i.e., the direction of the initial velocity is positive.

Example 10-7

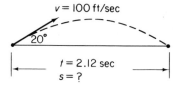

$v = 100$ ft/sec

$20°$

$t = 2.12$ sec
$s = ?$

Figure 10-11

A projectile is fired at an angle of 20° from the horizontal, with a velocity of 100 ft/sec. If it lands 2.12 sec later at the same elevation, how far did it travel horizontally?

From Figure 10-11, the horizontal component of the velocity is:

$$v_x = v\cos\theta$$
$$= 100\cos 20°$$
$$v_x = 94\,\text{ft/sec}$$
$$s = v_o t + \frac{1}{2}at^2$$
$$= 94 \times 2.12 + 0$$
$$s = 199\,\text{ft}$$

Example 10-8

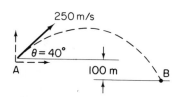

250 m/s

$\theta = 40°$

A   100 m   B

Figure 10-12

A projectile that is fired with an initial velocity of 250 m/s is inclined upward at an angle of 40°. It lands at a point 100 m lower than the initial point. Determine: (a) the time of flight, (b) the horizontal displacement; and (c) the final velocity at the point of landing.

Figure 10-12 shows the given information in sketch form.

(a) Between points A and B in the vertical direction, we have:

$$v_o = v\sin\theta = 250\sin 40°$$
$$v_o = 161\,\text{m/s}\uparrow$$
$$s = -100\,\text{m}$$
$$a = -9.81\,\text{m/s}^2$$
$$t = ?$$

Using Equation 10-3, we have:

$$s = v_o t + \frac{1}{2} a t^2$$

$$-100 = 161t + \frac{1}{2} \times (-9.81)t^2$$

$$4.9t^2 - 161t - 100 = 0$$

Using the quadratic equation, we have:

$$t = \frac{-b \pm \sqrt{b^2 - 4ac}}{2a}$$

$$= \frac{161 \pm \sqrt{(161)^2 - 4 \times 4.9 \times (-100)}}{2 \times 4.9}$$

$$= \frac{161 \pm 167}{9.8}$$

$$t = 33.4 \ (\text{or} -0.61) \, \text{s}$$

The projectile is in flight for 33.4 s.

(b) Between points A and B in the horizontal direction, we have:

$$v_o = v \cos \theta = 250 \cos 40°$$
$$v_o = 192 \, \text{m/s} \rightarrow$$
$$t = 33.4 \, \text{s}$$
$$a = 0$$
$$s = ?$$

Using Equation 10-3, we have:

$$s = v_o t + \frac{1}{2} a t^2$$

$$= 192 \times 33.4 + 0$$

$$s = 6400 \, \text{m}$$

The horizontal distance between A and B is 6400 m.

(c) The final velocity at B is the resultant of the vertical and horizontal velocities. The horizontal velocity remains constant $v_x = 192$ m/s.

Between A and B in the vertical direction, we get:

$$v_o = 161\,\text{m/s}\uparrow$$
$$v = v_y$$
$$a = -9.81\,\text{m/s}^2$$
$$t = 33.4\,\text{s}$$
$$v = v_o + at$$
$$v_y = 161 - 9.81 \times 33.4$$
$$= -167\,\text{m/s}\uparrow$$
$$v_y = 167\,\text{m/s}\downarrow$$

Adding v_x and v_y vectorially, we get:

$$\underline{v = 254\,\text{m/s} \;\sqrt{41°}}$$

Example 10-8 (Alternate Solution)

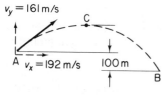

$v_y = 161\,\text{m/s}$

$v_x = 192\,\text{m/s}$ 100 m

Figure 10-13

To avoid using the quadratic equation, one could deal with a situation such as the one in Example 10-8 by analyzing the trajectory in small segments AC and CB shown in Figure 10-13. Point C is the top of the trajectory, the point at which velocity in the vertical direction is zero.

Between A and C in the vertical direction, we have:

$$v_o = 161\,\text{m/s}\uparrow$$
$$v = 0$$
$$a = -9.81\,\text{m/s}^2$$
$$t = ?$$
$$v = v_o + at$$
$$0 = 161 - 9.81\,t$$
$$t = 16.4\,\text{s}$$

Now solve for the vertical height of C above A.

$$s = v_o t + \frac{1}{2}at^2$$

$$= 161 \times 16.4 + \frac{1}{2}(-9.81)(16.4)^2$$

$$s = 1321\,\text{m}$$

Between C and B in the vertical direction, we have:

$$s = 1321 + 100 = 1421 \, m$$
$$a = -9.81 \, m/s^2$$
$$v_0 = 0$$
$$t = ?$$
$$s = v_0 t + \frac{1}{2} at^2$$
$$-1421 = 0 + \frac{1}{2} \times (-9.81)t^2$$
$$t = 17$$
$$\text{total time} = 16.4 + 17 = \underline{33.4 \, s}$$

The horizontal distance and the final velocity would be determined just as they were in the first solution.

PROBLEMS

10-1 A fork-lift truck lifts a pallet 2 m off the floor, moves 7 m ahead, and sets the pallet on a stack 1.5 m high. Determine the displacement of the pallet and the distance that it has travelled.

10-2 A conveyor 60 m long is inclined at an angle of 18° to the horizontal and deposits material 8 m below its top end. Determine the distance and displacement of material carried by this conveyor.

10-3 A dump truck hoists a box to an angle of 25° from the horizontal and pulls ahead 10 ft. Calculate the displacement of point A (Figure P10-3).

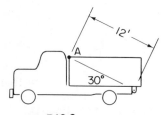

Figure P10-3

10-4 A new pipeline is filled with oil travelling at a constant velocity of 5 ft/sec. How long will it take after initial start-up before oil reaches a point 10 miles away?

10-5 A conveyor carries 0.25 m³ of gravel per meter of length. This conveyor dumps the gravel into a truck and can fill a 6 m³ truck in 10 sec. What is the speed of the conveyor?

10-6 A car travelling 40 mph is 140 ft from the near side of an intersection when the light turns yellow. The intersection is 80 ft wide and the light remains yellow for 4 seconds. If the car maintains its speed, will it be clear of the intersection when the light turns red? If so, with what distance to spare?

10-7 It takes a skier 5 minutes to go up the slope on a tow rope and to return to the bottom of the slope, a total distance of 700 m. Find the skier's average speed and average velocity.

10-8 A person follows a random path from A to B, a total distance of 30 m in 12 seconds. If the displacement between A and B is 15 m, find the average speed and the average velocity.

10-9 A stunt driver travels a total distance of 1172 ft moving at a constant speed of 40 mph while weaving through an obstacle course. If the total displacement of the car is 600 ft from the start of the course to the finish, determine the car's velocity.

10-10 A rocket initially at rest attains a velocity of 40 m/s in 5 seconds. What is its average acceleration?

10-11 A roller coaster car travelling at a constant velocity of 8 m/s goes around a 45° horizontal curve in 5 s. Determine the average acceleration.

10-12 A plane travelling 250 mph, while approaching the airport, banks into a 90° turn while maintaining its altitude and speed. Determine the average acceleration of the plane if it takes 50 sec. to complete the turn.

10-13 What is the deceleration rate of an elevator that comes to rest from a velocity of 180 m/min in a distance of 6 m?

10-14 A car travels 150 m in 10 s while constantly accelerating at 1 m/s². Determine the car's initial velocity and final velocity.

10-15 A car has uniform acceleration of 2 ft/sec² while travelling 390 ft in 15 sec. Determine the car's initial velocity and final velocity.

10-16   Car A, which is travelling at 100 km/h, is 30 m behind car B, which is travelling at 85 km/h. How far does car A travel in passing car B and reaching a point 40 m ahead of B? How long does it take?

10-17   An aircraft carrier with a 250 m long flight deck is travelling at 30 km/h into a 10 km/h wind. What is the acceleration of a plane that takes off into the wind from the carrier with an airspeed of 220 km/h? (Assume uniform acceleration.)

10-18   Car B stops at a traffic light; 900 ft ahead, car A stops at another traffic light. The light at A turns green, and 10 sec later the light at B is green. Car A accelerates uniformly at a rate of 3 ft/sec^2 to a constant velocity of 40 mph. Car B accelerates uniformly at a rate of 4 ft/sec^2 to a constant velocity of 60 mph. How long after car A starts out will it be overtaken by car B?

10-19   An object with negligible air resistance is dropped from a height of 60 m. Determine how long it takes to land and the velocity at which it lands.

10-20   A boy throws a stone vertically and it lands 5 seconds later. Assume that the stone landed at the same level as the boy's hand that released it. How high did he throw it ($g=9.81$ m/s^2)?

10-21   An astronaut jumps a distance of 3 ft vertically onto the surface of the moon ($g=5.31$ ft/sec^2). At what velocity does he hit the ground? What would be his velocity if he were subject to the earth's gravity ($g=32.2$ ft/sec^2)?

10-22   An object with an initial velocity of 25 m/s upward lands 80 m below its starting point. Find its maximum height, its total time in the air, and the velocity at the time of its landing.

10-23   A sky diver jumps from a balloon and falls freely for 6 sec. His parachute then opens decelerating him at a constant rate for 3 sec to a velocity of 18 ft/sec, which he maintains until he lands. (Neglect air resistance during the free fall.) If the sky diver jumped from an elevation of 6000 ft, determine: (a) his maximum velocity and (b) the total time elapsed during his travel from the plane to the ground.

10-24   A projectile is fired with an initial velocity of 40 m/s from a horizontal surface. The maximum height reached is 30 m. Determine the horizontal displacement and the firing angle with respect to the horizontal surface.

10-25 A golfer hits a ball, giving it a maximum height of 20 m. If the ball lands 130 m away at the level from which he hit it, what was its initial velocity? (Neglect air resistance.)

10-26 The golfer wants to hit the ball the same horizontal distance of 130 m and 20 m above his tee-off point as in Problem 10-21, but, in this case, the tee-off point is 10 m above the landing spot of the ball. What is the ball's initial velocity?

10-27 A projectile is fired from the top of a building as shown in Figure P10-27. Determine (a) the horizontal distance d and (b) its velocity just prior to striking point B.

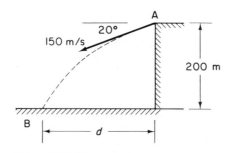

Figure P10-27

10-28 A ball is thrown with an initial velocity of 100 ft/sec and at an angle of 40° with the horizontal. Determine the minimum distance from a 50-ft high building from which the ball can be thrown and not hit the side of the building.

10-29 Will basketball player A in Figure P10-29 be able to throw the ball over player B and through the hoop without hitting the backboard or the ceiling? If so, determine the distance d that the ball will be from the ceiling at the ball's highest point. The ball is thrown with an initial velocity of 36 ft/sec at 50° to the horizontal. (Neglect the ball's diameter.)

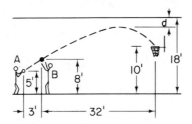

Figure P10-29

10-30 Grain pours from the end of a chute and lands in a container below (Figure P10-30). Determine the minimum velocity that it can have as it leaves the chute and still land in the container.

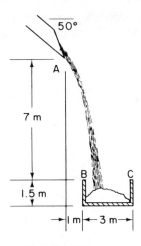

Figure P10-30

10-31 A projectile is fired at an angle of 30° to the horizontal up a slope that is at 10° to the horizontal. The firing velocity is 450 m/s. How far along the slope and with what velocity will the projectile strike the ground?

10-32 Water from a fire hose follows the path shown in Figure P10-32. Determine the velocity of the water as it leaves the hose.

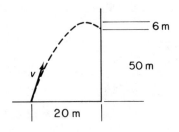

Figure P10-32

10-33   A bale-thrower attachment of a farm haybaler throws a completed bale into a wagon as shown in Figure P10-33. If $\theta = 48°$ and the velocity can be adjusted depending on the weight of the bale, what is the maximum velocity allowable before the bale will be thrown over the back of the trailer?

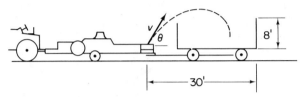

Figure P10-33

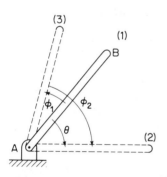

11

KINEMATICS–
ANGULAR MOTION

Consider the lever rotating about point A in Figure 11-1. As it rotates, it turns through some angle. For this reason, *rotational motion* of a body is also called *angular motion*. Let us therefore analyze the angular or rotational motion of lever AB in terms of displacement, distance, velocity, and acceleration.

Figure 11-1

11-1 ANGULAR DISPLACEMENT

The movement of lever AB is measured in terms of the angle through which it turns. It may turn a few degrees or several revolutions. A more useful unit used to measure angular displacement is the radian (rad). In Figure 11-1, AB turns through an angle of 1 *radian* when point B moves a distance on the circumference equal to radius AB. Since circumference is equal to $2\pi r$, there are 2π radians in one revolution. The radian is a dimensionless unit; it will be used later in Section 11-5 for conversion between rectilinear and angular motion. The angular displacement units are:

$$\text{one revolution} = 360 \text{ degrees} = 2\pi \text{ radians}$$

The symbol applied to angular displacement is the Greek letter theta (θ).

In Figure 11-1, suppose that lever AB moves from position (1) to position (2). The angular displacement is θ radians. If instead, the lever moved from position (1) to (3) and then to position (2), the angular displacement would still be θ radians, but the angular distance would be $(\phi_1 + \phi_2)$ radians. This distinction between displacement and distance is similar to the distinction that we made with rectilinear motion in Section 10-5.

Example 11-1

An ammeter needle, starting from a zero reading, deflects 40° clockwise and then returns 15° counterclockwise to indicate a final reading. What is its angular displacement and angular distance in terms of radians?

$$\text{angular displacement} = 40° - 15°$$
$$= 25° \text{ clockwise}$$

but

$$360° = 2\pi \text{ rad}$$
$$25° = \frac{25}{360} \times 2\pi$$

or

$$\theta = 0.437 \text{ rad } \downharpoonright$$
$$\text{Angular distance} = 40° + 15°$$
$$= \frac{55}{360} \times 2\pi$$
$$= 0.96 \text{ rad}$$

11-2 ANGULAR VELOCITY

Angular velocity is the rate of change of angular displacement and is represented by the Greek letter omega (ω). Like average rectilinear velocity, average angular velocity can be defined as:

$$\omega = \frac{\Delta\theta}{\Delta t} \tag{11-1}$$

where

$$\omega = \text{average angular velocity}$$
$$\theta = \text{angular displacement}$$
$$t = \text{time}$$

The most widely used units of ω are rad/s and rev/min (rpm).

Angular velocity, expressed in revolutions per minute (rpm), is not consistent with the SI metric system because the unit "minute" does not conform to the SI system and is classified as a "permitted" unit. Strictly speaking, the time units in the SI system are in seconds and multiples or submultiples, such as kiloseconds and milliseconds. Since the terms minute and hour are so universally used, the SI system is modified to permit their continued use.

Example 11-2

The crankshaft of an engine turns 800 revolutions clockwise in 4 minutes at constant velocity. What is its angular velocity in rad/s?

$$\omega = \frac{\Delta \theta}{\Delta t} = \frac{800 \text{ rev}}{4 \text{ min}}$$

$$= 200 \text{ rpm} \times \frac{2\pi \text{ rad/rev}}{60 \text{ s/min}}$$

$$\underline{\omega = 20.9 \text{ rad/s} \downarrow}$$

11-3   ANGULAR ACCELERATION

Angular acceleration is the rate of change of angular velocity and is represented by the Greek letter alpha (α). Like rectilinear motion, angular acceleration can be defined as:

$$\alpha = \frac{\Delta \omega}{\Delta t} \qquad (11\text{-}2)$$

where

$$\alpha = \text{average angular acceleration}$$

The usual units of α are rad/s^2.

Example 11-3

A flywheel accelerates uniformly from rest to a speed of 20 rpm counter-clockwise in 5 seconds. What is the average angular acceleration in rad/s²?

$$\Delta\omega = 20 \text{ rpm}$$

$$\Delta\omega = 20 \times \frac{2\pi}{60} \text{ rad/s}$$

$$\alpha = \frac{\Delta\omega}{\Delta t}$$

$$= \frac{20}{5} \times \frac{2\pi}{60}$$

$$\underline{\alpha = 0.42 \text{ rad/s}^2 \curvearrowleft}$$

11-4   ANGULAR MOTION WITH UNIFORM ACCELERATION

In Section 10-5, we set forth three equations for rectilinear motion with uniform acceleration. These three equations are repeated in Table 11-1 with their analogous angular motion equations. The angular terms θ, ω, and α are simply used in the original rectilinear equations. Keep in mind that all of these equations are based on uniform acceleration.

Table 11-1

RECTILINEAR	ANGULAR	
$s = v_0 t + \frac{1}{2}at^2$	$\theta = \omega_0 t + \frac{1}{2}\alpha t^2$	(11-3)
$v = v_0 + at$	$\omega = \omega_0 + \alpha t$	(11-4)
$v^2 = v_0^2 + 2as$	$\omega^2 = \omega_0^2 + 2\alpha\theta$	(11-5)

In common units

$$\theta = \text{displacement in rad}$$
$$\omega_0 = \text{initial velocity in rad/s}$$
$$\omega = \text{final velocity in rad/s}$$
$$\alpha = \text{uniform acceleration in rad/s}^2$$
$$t = \text{time in seconds}$$

Example 11-4

When a small plane touches down on a landing strip, its wheels accelerate from rest to a speed of 1100 rpm in three seconds. Calculate (a) the average angular acceleration; and (b) the number of revolutions of each wheel in this time period.

(a) the given information is:

$$\omega_0 = 0$$

$$\omega = 1100 \times \frac{2\pi}{60} = 115 \text{ rad/s}$$

$$t = 3 \text{ s}$$

$$\alpha = ?$$

$$\omega = \omega_0 + \alpha t$$

$$115 = 0 + \alpha \times 3$$

$$\underline{\alpha = 38.3 \text{ rad/s}^2}$$

(b)

$$\theta = \omega_0 t + \frac{1}{2}\alpha t^2$$

$$= 0 + \frac{1}{2} \times 38.3 \times (3)^2$$

$$= 172 \text{ rad}$$

$$= \frac{172}{2\pi} \text{ rev}$$

$$\underline{\theta = 27.4 \text{ rev}}$$

Example 11-5

A propeller fan 3 m in diameter used in a cooling tower comes to rest with uniform acceleration from a clockwise speed of 600 rpm. If it turns through 15 revolutions while stopping, calculate the time that it requires to stop.

$$\omega_0 = 600 \text{ rpm} = \frac{600 \times 2\pi}{60} = 62.8 \text{ rad/s}$$

$$\omega = 0$$

$$\theta = 15 \text{ rev} = 15 \times 2\pi = 94.2 \text{ rad}$$

The deceleration α must be found first.

$$\omega^2 = \omega_0^2 + 2\alpha\theta$$
$$0 = (62.8)^2 + 2\alpha \times 94.2$$
$$-3940 = 188.4\alpha$$
$$\alpha = -21 \text{ rad/s}^2 \text{ clockwise}$$

or

$$\alpha = 21 \text{ rad/s}^2 \text{ counterclockwise}$$

The minus sign indicates that there is deceleration in the direction of the rotation.

Using Equation 11-4 to calculate time, we get:

$$\omega = \omega_0 + \alpha t$$
$$0 = 62.8 - 21t$$
$$\underline{t = 3 \text{ s}}$$

11-5   RELATIONSHIP BETWEEN RECTILINEAR AND ANGULAR MOTION

There are many situations in which rectilinear and angular or rotational motion are combined. Belts on pulleys and revolving car wheels are two such examples. We must be able to convert between rectilinear values (s, v, a) and angular values (θ, ω, α).

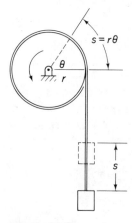

A hoist drum with a cable wound around it can rotate and lift a weight (Figure 11-2). If the drum turns counterclockwise through an angle of one radian, then the amount of cable wound onto the drum is equal to the radius; the weight is also lifted a distance s, equal to the radius. For two radians of drum rotation, the cable wound onto the drum equals $2r$ and distance $s = 2r$. Distance s depends on the radius and amount of rotation measured in radians or, in equation form:

$$s = r\theta \qquad (11\text{-}6)$$

Figure 11-2

where s and r must have the same units,

such as meters or feet; θ is expressed in radians.

Dividing both sides by t, we have:

$$\frac{s}{t} = r\frac{\theta}{t}$$

$$v = r\omega \qquad (11\text{-}7)$$

Dividing both sides by t again, we have:

$$\frac{v}{t} = r\frac{\omega}{t}$$

$$a = r\alpha \qquad (11\text{-}8)$$

A feature common to all three equations is that a rectilinear value equals the radius multiplied by the angular value.

Example 11-6

Wire for a transmission line is unrolled from a reel at a rate of 750 ft/min. The radius of the reel is 2.5 ft for the instant considered. What is the angular speed of the reel in rpm?

$$v = r\omega$$
$$750 = 2.5\omega$$

$$\omega = 300 \text{ rad/min}$$

$$= \frac{300}{2\pi} \text{ rev/min}$$

$$\underline{\omega = 47.8 \text{ rpm}}$$

Example 11-7

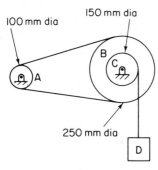

100 mm dia

150 mm dia

250 mm dia

As shown in Figure 11-3, weight D is suspended by a rope wrapped around pulley C. Pulleys B and C are fastened together and pulley A is belt-driven by pulley B. Starting from rest, weight D drops 18 m in three seconds. For each pulley, determine (a) the number of revolutions; (b) the angular velocity, and (c) the angular acceleration at $t = 3$ s.

Figure 11-3

Calculate the rectilinear values before converting to angular values.

$$s = 18 \text{ m}$$
$$t = 3 \text{ s}$$
$$v_0 = 0$$
$$v = ?$$
$$a = ?$$
$$s = v_0 t + \frac{1}{2} a t^2$$
$$18 = 0 + \frac{1}{2} \times a \times (3)^2$$
$$a = 4 \text{ m/s}^2 \downarrow$$
$$v = v_0 + at$$
$$= 0 + (4 \times 3)$$
$$v = 12 \text{ m/s} \downarrow$$

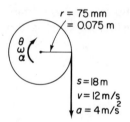

$r = 75 \text{ mm}$
$= 0.075 \text{ m}$

$s = 18 \text{ m}$
$v = 12 \text{ m/s}$
$a = 4 \text{ m/s}^2$

Figure 11-4

Considering pulley C first (Figure 11-4), we have:

$$s = r\theta$$
$$18 = 0.075\theta$$
$$\theta = 240 \text{ rad}$$
$$\underline{\theta = 38.2 \text{ rev} \searrow}$$
$$v = r\omega$$
$$12 = 0.075\omega$$
$$\underline{\omega = 160 \text{ rad/s} \searrow}$$
$$a = r\alpha$$
$$4 = 0.075\alpha$$
$$\underline{\alpha = 53.3 \text{ rad/s}^2 \searrow}$$

Since pulley B is fastened to pulley C, it must have the same angular values as pulley C; that is:

$$\underline{\theta = 38.2 \text{ rev} \searrow}$$
$$\underline{\omega = 160 \text{ rad/s} \searrow}$$
$$\underline{\alpha = 53.3 \text{ rad/s}^2 \searrow}$$

The belt has the same speed throughout its length; therefore, the tangential velocity of A equals the tangential velocity of B.

$$v_A = v_B$$
$$r_A \omega_A = r_B \omega_B$$
$$\omega_A = \frac{r_B}{r_A} \times \omega_B$$
$$= \frac{0.125}{0.050} \times 160$$
$$\underline{\omega_A = 400 \text{ rad/s } \downarrow}$$

Angular displacement and acceleration can be found by means of the same ratio of radii.

$$\theta_A = \frac{r_B}{r_A} \theta_B$$
$$= \frac{0.125}{0.050} \times 38.2$$
$$\underline{\theta_A = 95.5 \text{ rev } \downarrow}$$
$$\alpha_A = \frac{0.125}{0.050} \times 53.3$$
$$\underline{\alpha_A = 133 \text{ rad/s}^2 \downarrow}$$

11-6   NORMAL AND TANGENTIAL ACCELERATION

As we demonstrated in Section 11-5, a point on the rim of a wheel—with radius r and angular acceleration α—has a rectilinear acceleration of $r\alpha$. The direction of the motion is tangent to the radius. We will place special emphasis on *tangential acceleration* by designating it a_t; since there is another acceleration present, normal acceleration, which we will designate a_n.

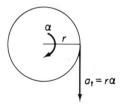

Figure 11-5

Tangential acceleration a_t is due to the wheel's speed changing, i.e., changing *magnitude* of its v and ω values. Another acceleration, *normal acceleration*, is present even with constant wheel speed. Normal acceleration is due to the change in *direction* of velocity, as was discussed in Section 10-4.

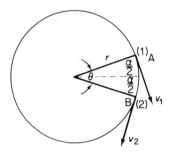

Figure 11-6

In Figure 11-6, a wheel turns at constant speed, and a point on the rim travels from point A to point B. The tangential velocities, v_1 and v_2, are equal in magnitude.

normal acceleration = velocity change (in direction)

$$a_n = \Delta v$$
$$= v_2 - v_1$$
$$a_n = v_2 + (-v_1)$$

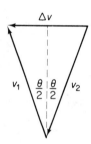

Figure 11-7

Figure 11-7 shows the vector addition of $v_2 + (-v_1)$ to obtain Δv. From this vector triangle, we have:

$$\sin\frac{\theta}{2} = \frac{\Delta v/2}{v}$$

where

$$v = v_2 \text{ or } v_1$$
$$\Delta v = 2v\sin\frac{\theta}{2}$$

For small angles:

$$\sin\frac{\theta}{2}\text{degrees} = \frac{\theta}{2}\text{radians}$$

$$\Delta v = 2v\frac{\theta}{2} = v\theta \qquad (11\text{-}9)$$

For small angles:

$$\text{chord length AB} = \text{arc length AB}$$
$$\Delta v = r\theta$$
$$\text{distance} = \text{velocity} \times \text{time}$$

or

$$AB = v \times \Delta t$$

and therefore:

$$v \times \Delta t = r\theta$$

$$\Delta t = \frac{r\theta}{v} \tag{11-10}$$

Using a basic equation (Equation 10-2), we have:

$$a = \frac{\Delta v}{\Delta t}$$

and substituting (from Equations 11-9 and 11-10), we get:

$$a = \frac{v\theta}{\dfrac{r\theta}{v}}$$

or, for normal acceleration:

$$a_n = \frac{v^2}{r} \tag{11-11}$$

The direction of a_n must be the same as that of Δv, which was along the radius toward the center of rotation.

Since $v = r\omega$, another version of the equation for normal acceleration is:

$$a_n = \frac{r^2\omega^2}{r}$$

or

$$a_n = \omega^2 r \tag{11-12}$$

The reason this characteristic is called normal acceleration is that it acts inwardly along the radius, i.e., at right angles to—or normal to—the circular path at that instant.

Since there may be both normal and tangential acceleration simultaneously, the total acceleration would have to be the vector sum of a_n and a_t. The following examples will illustrate this.

Example 11-8

Determine the normal acceleration of a car travelling around a circle with a radius of 200 ft at a speed of 30 mph. (60 mph = 88 ft/sec)

$$v = 30 \text{ mph} = \frac{30}{60} \times 88 = 44 \text{ ft/sec}$$

$$a_n = \frac{v^2}{r}$$

$$= \frac{(44)^2}{200}$$

$$a_n = 9.68 \text{ ft/sec}^2$$

Example 11-9

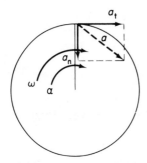

Figure 11-8

A pulley with a radius of 80 mm is turning at a velocity of 150 rpm clockwise and is accelerating at 100 rad/s². Determine the total acceleration of a point on the rim at the instant during which it is at the top of the pulley (Figure 11-8).

$$\omega = 150 \text{ rpm} = \frac{150 \times 2\pi}{60} = 15.7 \text{ rad/s}$$

$$a_n = \omega^2 r$$

$$= (15.7)^2 \times 0.080$$

$$a_n = 19.7 \text{ m/s}^2 \downarrow$$

$$a_t = r\alpha$$

$$= 0.08 \times 100$$

$$a_t = 8 \text{ m/s}^2 \rightarrow$$

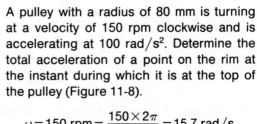

The total acceleration a is the sum of a_n and a_t.

$$a = \sqrt{(8)^2 + (19.7)^2} \qquad \text{(Figure 11-9)}$$

$$\underline{a = 21.3 \text{ m/s}^2 \searrow^{67.9°}}$$

$$\tan\theta = \frac{19.7}{8}$$

$$\theta = 67.9°$$

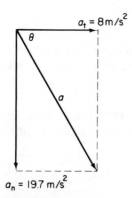

Figure 11-9

PROBLEMS

11-1 The minute hand of a clock travels from the 12 to the 9, where-upon it is discovered that the clock is fast. The minute hand is turned back and now points at the 6. What is the angular distance and displacement of the hand in radians?

11-2 Convert 4 radians to degrees.

11-3 Convert 15 revolutions to radians.

11-4 Convert 1480 degrees to (a) revolutions and (b) radians.

11-5 Convert 90 rpm to radians per second.

11-6 A wheel turns 80 revolutions in 3 minutes. What is its angular speed in rad/min?

11-7 It is calculated that a fan may be rotated at 220 rad/s before stress limits are exceeded. What is this speed in rpm?

11-8 Rod A in Figure P11-8 reciprocates up and down 50 times in 40 s due to the action of the rotating cam B. Determine the angular speed of cam B in rad/min.

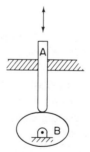

Figure P11-8

11-9 Convert 5 rad/s² to (a) rad/min² and (b) rev/min².

11-10 Starting from rest, a shaft accelerates uniformly to 800 rpm in 10 seconds. Determine the angular acceleration in rad/s².

11-11 An angular velocity is increased from 100 rad/min to 250 rad/min in 20 seconds. Determine the uniform angular acceleration in rad/s².

11-12 A rotating drive shaft decelerates uniformly from 900 rpm to 650 rpm in 6 s. Determine the angular deceleration and the total number of revolutions in the 6-second time interval.

11-13 A wheel turns through 500 revolutions while accelerating from 80 to 110 rad/s. Determine the angular acceleration and the time required.

11-14 A revolving vane anemometer turns 40 revolutions in 30 s at constant velocity. Determine angular velocity in (a) rpm and (b) rad/s.

11-15 A pulley rotating at 80 rpm clockwise changes its rotation to 120 rpm counterclockwise in 3 seconds. Determine (a) angular deceleration and (b) the total number of revolutions in the 3 s time period.

11-16 When a power lawn mower is stalled, its blade is brought to rest from a speed of 2000 rpm, turning 250 revolutions in the process. Determine (a) the angular deceleration and (b) the time required for the lawn mower to stop.

11-17 A bicycle wheel held up off the ground is spun at 350 rpm and is then allowed to coast to a stop while turning through 4000 revolutions. How long does it take to stop? (Assume constant deceleration.)

11-18 A flywheel accelerates for 8 s at 1.3 rad/s^2 from a speed of 40 rpm. Determine (a) the total number of revolutions and (b) the final angular speed.

11-19 A large flywheel used on a punch press is slowed from 70 rpm to 65 rpm in 2 s. Determine its rate of deceleration.

11-20 A hand-started motor is accelerated from rest to 150 rpm in 1 s when the rope is pulled. It then starts and, under load, accelerates uniformly to 3600 rpm in 7 seconds. Determine (a) the angular acceleration in each case and (b) the total number of revolutions of the motor.

11-21 A large saw blade in a sawmill accelerates from rest to 800 rpm while turning 160 revolutions. Determine the time and the angular acceleration.

11-22 Starting from rest, a water turbine turns 130 revolutions in accelerating uniformly to its operating speed of 200 rpm. Determine (a) the angular acceleration; (b) the total time required; and (c) the turbine speed after the first 40 s.

11-23 A drive pulley decelerates at 5 rad/s² from 300 rpm to 180 rpm. It then accelerates to 260 rpm in 2 s. Determine (a) the total time required; (b) the acceleration rate; and (c) the total number of revolutions.

11-24 Starting from rest, a wheel accelerates uniformly to 300 rpm in 5 seconds. After rotating at 300 rpm for some time, it is braked to a stop in 90 seconds. If the total number of revolutions of the wheel is 800, calculate the total time of rotation.

11-25 Rocker arms AB and CD in Figure P11-25 rock the horizontal deck of a combine back and forth. Assume that each arm accelerates uniformly to the midpoint of its arc and then decelerates uniformly to the end of the arc. A complete cycle takes 2 s. Determine (a) the maximum angular velocity and (b) the angular acceleration.

Figure P11-25

11-26 Determine the length of an arc that has a radius of 50 m and an included angle of 2.5 rad.

11-27 What is the length of a curve that has a radius of 600 m and passes through an angle of 1.4 radians?

11-28 An automatic pipewelder can weld 40 in./min. How long does it take to complete one pass around a pipe 4 ft in diameter?

11-29 By means of a stroboscope, the speed of a pulley 200 mm in diameter is found to be 1600 rpm. Determine the speed of a belt passing over this pulley.

11-30 A cylinder 2 ft in diameter is rolled along a flat surface. If it turns 4 revolutions, how far has it moved along the flat surface?

11-31 The rear wheels of a tractor are 1.8 m in diameter, and the front wheels are 0.8 m. If the rear wheels are turning at 40 rpm determine (a) the velocity of the tractor, and (b) the angular speed of the front wheels. Assume no slipping.

11-32 If the rocker arms in Problem 11-25 have radii of 90 mm, determine the maximum horizontal velocity of the deck.

11-33 If a spotlight rotating in a horizontal plane at 2 rpm is located 130 m from you, at what speed would the light beam flash across you?

11-34 In the manufacture of sheet steel, the sheet, which has a velocity of 12 m/s, is drawn between two rollers. If the rollers are 180 mm in diameter, determine their speed of rotation in rpm.

11-35 A pulley 10 in. in diameter is belt-driven by a pulley 6 in. in diameter. A pulley 4 in. in diameter is used for the belt tightener. If the 10-in. diameter pulley is turning at 120 rpm, determine the belt speed and the rpm of the other two pulleys.

11-36 A hand sling-psychrometer is composed of two thermometers with a handle at one end and a wet wick at the other. By whirling the thermometers about the handle, one obtains a relative humidity reading. If the distance from the handle to the wick is 400 mm and the wick must have a velocity of 5 m/s for a reliable reading, at what rpm must the thermometers be whirled?

11-37 When a car owner has the standard 25-in. diameter wheels on his car, the speedometer indicates a true 60 mph. What is the wheel rpm at this speed? If he now installs wheels 26 in. in diameter and drives at a speed so that the speedometer indicates 60 mph, what is the wheel rpm and at what speed is he actually travelling?

11-38 The trailer shown in Figure P11-38 has a pulley C fastened to wheel B. Drum A is belt driven by pulley C. If the trailer is pulled with a velocity of 9 km/h, determine the angular velocity of drum A.

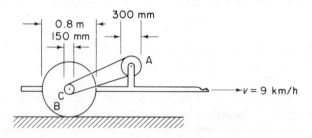

Figure P11-38

11-39 The flywheel in Figure P11-39 is decelerating at 3 rad/s² while turning at 12 rad/s clockwise. Determine the tangential acceleration and velocity of point P on the rim.

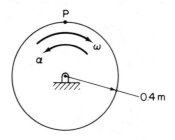

Figure P11-39

11-40 A lathe chuck 8 in. in diameter accelerates uniformly from rest to 400 rpm in 1 second. What is the tangential acceleration of a point on the rim?

11-41 Pulleys C and D in Figure P11-41 are fastened together. Weights A and B are supported by ropes wound around the pulleys as shown. If weight A is accelerating downward at 2 m/s², determine (a) the angular acceleration of pulleys C and D and (b) the linear acceleration of weight B.

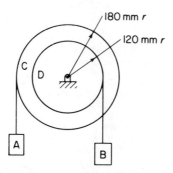

Figure P11-41

11-42 In the pulley system shown in Figure P11-42, $\alpha = 2$ rad/s² clockwise and $\omega = 8$ rad/s clockwise for pulley A. Determine (a) angular velocity of all pulleys; (b) angular acceleration of all pulleys; and (c) tangential acceleration of point P.

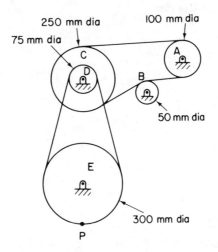

Figure P11-42

11-43 Weight D in Figure P11-43 is lifted from rest at constant acceleration to a height of 50 ft in 20 sec by means of a cable wound around drum A. Assume that the diameter of A remains constant and determine (a) the angular distance travelled by gears B and C; (b) the angular acceleration of gears B and C; and (c) the tangential acceleration of the point of contact between gears B and C. (Number of gear teeth shown is not representative.)

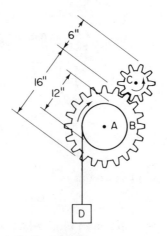

Figure P11-43

11-44   What is the normal acceleration of a point 9 in. from the center of a rotor that is turning at 4500 rpm?

11-45   A point 150 mm from the center of rotation has a normal acceleration of 4000 m/s². Determine the speed of rotation.

11-46   A weight is whirled on the end of a 2.2 m rope. The rope has an angular speed of 600 rad/min. Determine the normal acceleration of the weight.

11-47   A plane, while maintaining its altitude, turns in an arc with a radius of 900 m. If the plane has a velocity of 410 km/h, what is the normal acceleration of the plane?

11-48   A boy on a swing with 20-ft ropes has a velocity of 15 ft/sec at his lowest point. What is his normal acceleration at this point?

11-49   A highway curve with a radius of 1200 ft is designed in such a way that it is unsafe for a car on this curve to have a normal acceleration greater than 5 ft/sec². What is the maximum velocity that a car may attain safely on this curve?

11-50   Determine the normal and tangential accelerations for point P in Figure P11-50.

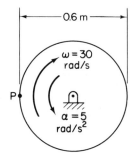

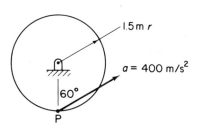

Figure P11-50   **Figure P11-52**

11-51   A car accelerates on a 500 m radius curve at 1.2 m/s². What is the speed of the car when it has a total acceleration of 2 m/s²?

11-52   Point P on the rim of the wheel in Figure P11-52 has a total acceleration of 400 m/s². Determine the angular velocity and angular acceleration of the wheel.

11-53   A car accelerates uniformly from rest to 50 mph over a distance of 1/8 mile along a curve of 700 ft radius. Determine the total acceleration at the instant that the speed of 50 mph is reached.

11-54   In an amusement ride, people sit in a bullet-shaped compartment, such as A in Figure P11-54, and are rotated in a vertical circle. Determine the normal and tangential accelerations due to the rotation for the position shown and $t = 10$ s. (The ride accelerates from rest to 3 rad/s in 10 s.

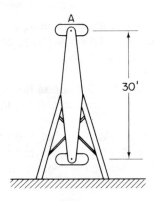

Figure P11-54

PLANE MOTION

12-1 RELATIVE MOTION

When a body is in motion, it may have displacement, velocity, and acceleration. The following discussion of relative motion applies to all three of these variables; for easier visualization, we will begin with velocity.

A common assumption that is often made about velocity is that velocity is stated with respect to earth. The reason for this common reference is that the earth appears stationary to us. For most engineering calculations, this is a safe assumption, and we will use it here. Since the earth is considered stationary, a velocity measured with respect to earth is an *absolute velocity*.

Relative velocity comes into play when the velocity of one object is related to that of another reference object that is also moving. For example, if car A were travelling at 40 km/h (absolute) and were passed by car B travelling at 60 km/h (absolute), car B would have a velocity of 20 km/h relative to car A. Car B, therefore, would have an absolute velocity of 60 km/h and a relative velocity of 20 km/h with respect to the car A. As you can appreciate, a relative velocity has no meaning unless the reference or point to which the velocity is relative is stated.

Since we are concerned with the velocities of cars A and B and the velocity of B with respect to A, the following notation will be used for absolute and relative velocity:

$$v_A = \text{velocity of A}$$
$$v_B = \text{velocity of B}$$
$$v_{B/A} = \text{velocity of B with respect to A}$$

For the situation in our example above (Figure 12-1):

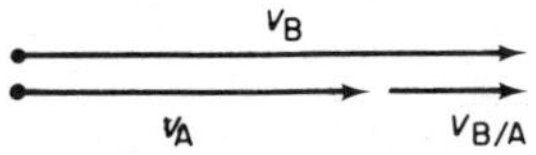

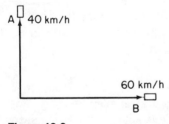

$$v_B = v_A + v_{B/A}$$
$$60 = 40 + 20$$
$$60 = 60$$

Figure 12-1

Keep in mind that we could have also written:

$$s_B = s_A + s_{B/A}$$
$$a_B = a_A + a_{B/A}$$

When the velocities are not in the same direction, but on some angle to one another, we must employ vector addition—possibly using the sine law.

Example 12-1

A ↕ 40 km/h

60 km/h
B

Figure 12-2

Starting from the same point, car A travels north at 40 km/h and car B travels east at 60 km/h. Determine the velocity of B with respect to A.

To visualize the velocity of B with respect to that of A, you would have to be in car A looking at car B (Figure 12-2).

Our previous equation applies, but directions must be shown as well.

$$v_B = v_A + v_{B/A}$$

$$\overrightarrow{60 \text{ km/h}} = \overset{\uparrow}{40 \text{ km/h}} + v_{B/A}$$

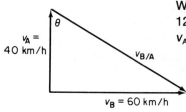

We construct the vector triangle (Figure 12-3) as indicated by the equation; that is, v_A and $v_{B/A}$ are tip-to-tail.

$$v_{B/A} = \sqrt{1600 + 3600}$$

$$\underline{v_{B/A} = 72 \text{ km/h S } 56.3° \text{ E}}$$

$$\tan \theta = \frac{60}{40}$$

$$\theta = 56.3°$$

Figure 12-3

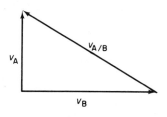

The velocity of A with respect to B ($v_{A/B}$) is equal and opposite to $v_{B/A}$ and could have been solved for by means of the vector triangle shown in Figure 12-4 and the following equation.

$$v_A = v_B + v_{A/B}$$

$$\overset{\uparrow}{40 \text{ km/h}} = \overrightarrow{60 \text{ km/h}} + v_{A/B}$$

Figure 12-4

Note that a way of ensuring that you have the correct equation is to check that the subscripts on the right side of the equation cancel out to equal the subscript on the left.

$$v_A = v_B + v_{A/B}$$

$$v_A = v_A$$

Now that we have looked at relative velocity, or motion between two separate objects, let us investigate the relative motion between two points on the same object. This will occur when an object moves with general plane motion consisting of simultaneous translation and rotation.

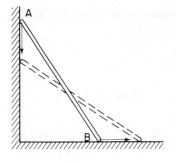

Figure 12-5

An example of plane motion is a bar leaning against a wall, as in Figure 12-5; the bottom of the bar is slipping to the right. The *translational motion* consists of A moving downward and B moving to the right. *Rotational motion* is also evident as the bar rotates in a counterclockwise direction about its center. The complete plane motion can be analyzed and divided into its individual components: translational motion and then rotational (Figure 12-6); or rotational motion and then translational (Figure 12-7).

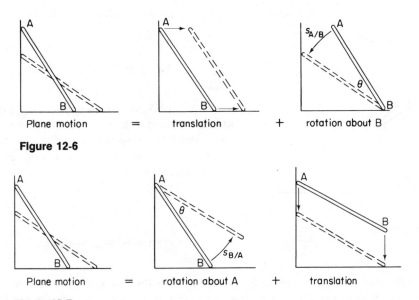

| Plane motion | = | translation | + | rotation about B |

Figure 12-6

| Plane motion | = | rotation about A | + | translation |

Figure 12-7

In Figure 12-6, the displacement of A with respect to B, $s_{A/B}$, is the arc of a circle or:

$$s_{A/B} = r\theta$$
$$= (AB)\theta$$

Similarly, in Figure 12-7:

$$s_{B/A} = (AB)\theta$$

The equation describing the plane motion would be:

$$s_A = s_B + s_{A/B}$$
$$\downarrow = \rightarrow + \swarrow$$

or

$$s_B = s_A + s_{B/A}$$
$$\rightarrow = \downarrow + \nearrow$$

Example 12-2

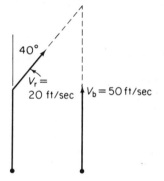

Figure 12-8

A football receiver runs straight downfield (north), turns 40° to his right, and, maintaining his speed of 20 ft/sec, catches a football that is travelling at 50 ft/sec due north (Figure 12-8). Determine the velocity of the ball with respect to the receiver.

Use the following notation.

$$v_r = \text{velocity of the receiver}$$
$$v_b = \text{velocity of the ball}$$
$$v_{b/r} = \text{velocity of the ball with respect to the receiver}$$

$$v_b = v_r + v_{b/r}$$
$$v_{b/r} = v_b - v_r$$

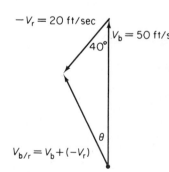

Figure 12-9

Constructing the vector triangle (Figure 12-9) and applying the cosine law; we get:

$$(v_{b/r})^2 = 20^2 + 50^2 - 2$$
$$\times (20) \times 50 \cos 40°$$
$$v_{b/r} = 37 \text{ ft/sec} \nwarrow^{20.3°}|$$

$$\frac{20}{\sin\theta} = \frac{37}{\sin 40°}$$
$$\theta = 20.3°$$

Example 12-3

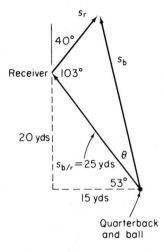

Figure 12-10

Suppose now that the quarterback and receiver are located as in Figure 12-10. As before, the receiver and the ball have velocities of 20 ft/sec and 50 ft/sec, respectively, and the ball is released the instant the receiver makes his 40° turn. Determine (a) the distance travelled by the ball before it is caught and (b) the angle θ at which the quarterback must lead his receiver.

Distances s_r and s_b are covered in the same time, so:

$$t = \frac{s_r}{v_r} = \frac{s_b}{v_b}$$

$$\frac{s_r}{20} = \frac{s_b}{50}$$

$$s_r = 0.4 \, s_b$$

By the relative velocity formula and the vector triangle, we have:

$$s_b = s_r + s_{b/r}$$

Using the cosine law, we obtain:

$$(s_b)^2 = (s_r)^2 + (s_{b/r})^2 - 2 s_r s_{b/r} \cos 103°$$
$$(s_b)^2 = (0.4 s_b)^2 + (25)^2 - 2 \times 0.4 s_b \times 25 (-0.225)$$
$$s_b^2 - 5.36 s_b - 745 = 0$$

Solving by means of the quadratic equation, we obtain:

$$s_b = 30 \text{ yd}$$
$$s_r = 0.4 s_b$$
$$= 0.4 \times 30$$
$$s_r = 12 \text{ yd}$$

Using the sine law, we obtain:

$$\frac{s_b}{\sin 103°} = \frac{s_r}{\sin \theta}$$

$$\sin \theta = \frac{s_r}{s_b} \sin 103°$$

$$\sin \theta = 0.4 \times 0.975$$

$$\underline{\theta = 23°}$$

The ball travels 30 yd, and the quarterback leads the receiver by 23°.

Example 12-4

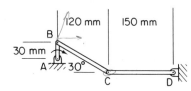

Figure 12-11

Bar AB of the linkage shown in Figure 12-11 rotates at 20 rad/s clockwise. For the position shown, determine (a) the velocity of pin C and (b) the angular velocity of link CD.

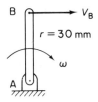

Figure 12-12

Considering bar AB (Figure 12-12), we have:

$$v_B = r\omega$$

$$= 30 \times 20$$

$$v_B = 600 \text{ mm/s} \rightarrow$$

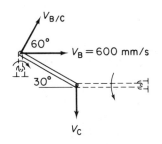

Figure 12-13

For bar BC (Figure 12-13), we can write the equation:

$$v_B = v_C + v_{B/C}$$

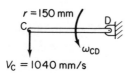

$V_B = 600$ mm/s

V_C $V_{B/C}$

$30°$

$60°$

Figure 12-14

The vector triangle is drawn according to the equation; that is, v_C and $v_{B/C}$ are tip-to-tail as in Figure 12-14.

$$\tan 30° = \frac{v_B}{v_C}$$

$$v_C = \frac{600}{0.577}$$

$$= 1040 \text{ mm/s}\downarrow$$

$$\underline{v_C = 1.04 \text{ m/s}\downarrow}$$

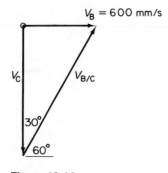

$r = 150$ mm

C ———————— D

ω_{CD}

$V_C = 1040$ mm/s

Figure 12-15

From Figure 12-15, we have:

$$v_C = r\omega_{CD}$$

$$\omega_{CD} = \frac{1040}{150}$$

$$\underline{\omega_{CD} = 6.93 \text{ rad/s } \circlearrowleft}$$

Example 12-5

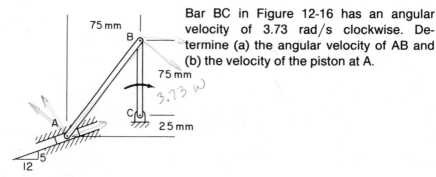

75 mm

B

75 mm

3.73 ω

C

25 mm

A

5

12

Figure 12-16

Bar BC in Figure 12-16 has an angular velocity of 3.73 rad/s clockwise. Determine (a) the angular velocity of AB and (b) the velocity of the piston at A.

V_B

$r = 75$ mm

$\omega = 3.73$ rad/s

Figure 12-17

$$v_B = r\omega \qquad \text{(Figure 12-17)}$$

$$= 75 \times 3.73$$

$$v_B = 280 \text{ mm/s} \rightarrow$$

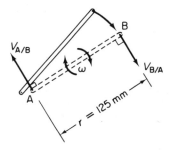

Figure 12-18

Considering member AB in its position in Figure 12-18, we can see that, although it has translational motion, it rotates clockwise. The relative velocity between points A and B, which causes this rotation, can be expressed either as $v_{A/B}$ or as $v_{B/A}$. Both velocities must be at right angles to bar AB since they are tangential to radius arm AB. Either velocity may be used in our equation; let us arbitrarily use $v_{A/B}$.

$$v_A = v_{A/B} + v_B$$

$$\angle_5 = {}_3\nwarrow + 280 \overrightarrow{\text{mm/s}}$$

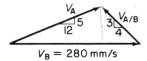

Figure 12-19

After constructing the vector triangle (Figure 12-19), we can calculate the internal angles of the triangle and use the sine law; alternatively, we can use the method described below.

Show the vertical components of the slopes as a common denominator 15 (Figure 12-20) with correspondingly larger horizontal components.

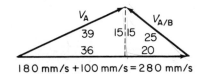

Figure 12-20

$$v_A = \frac{39}{36} \times 180$$

$$v_A = 195 \text{ mm/s} \ \angle^5_{12}$$

$$v_{A/B} = \frac{25}{20} \times 100$$

$$= 125 \text{ mm/s} \ {}^3\nwarrow_4$$

$$v_{A/B} = (AB)\omega_{AB}$$

$$125 = 125\omega_{AB}$$

$$\underline{\omega_{AB} = 1 \text{ rad/s} \ \circlearrowright}$$

As was stated above, relative acceleration can be handled by using the type of equation that was used for distance and velocity; for example:

$$a_B = a_{B/A} + a_A$$

When points A and B are on the same object, such as a link of a mechanism, each acceleration term in the above equation will consist of two components, a *tangential component* and a *normal component*. We are now dealing with six acceleration values rather than just three, as the equation may indicate. This is illustrated in the following example.

Example 12-6

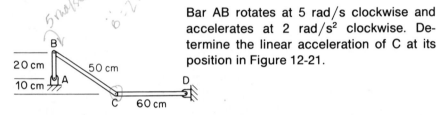

Bar AB rotates at 5 rad/s clockwise and accelerates at 2 rad/s² clockwise. Determine the linear acceleration of C at its position in Figure 12-21.

Figure 12-21

Figure 12-22 shows the equation to be used as well as vector sketches of normal and tangential acceleration in all cases.

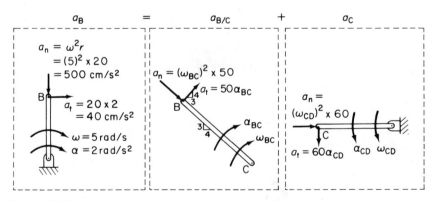

Figure 12-22

From Figure 12-22, we can see that the unknown velocities, ω_{BC} and ω_{CD}, must be found before any normal acceleration values can be determined. Figure 12-23 shows the relative velocity equation used in solving for these velocities.

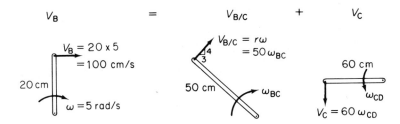

$$V_B \qquad = \qquad V_{B/C} \qquad + \qquad V_C$$

Figure 12-23

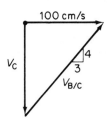

Figure 12-24

From the vector triangle of Figure 12-24, we have:

$$V_C = \frac{4}{3} \times 100 = 133 \text{ cm/s}$$

$$V_{B/C} = \frac{5}{3} \times 100 = 167 \text{ cm/s}$$

$$\omega_{CD} = \frac{133}{60} = 2.22 \text{ rad/s}$$

$$\omega_{BC} = \frac{167}{50} = 3.34 \text{ rad/s}$$

Return now to the acceleration equation in Figure 12-22 with the angular velocity values calculated above.

Note that the equation shows two vectors (a_n and a_t) of B equal to the sum of four other vectors. Now write an equation showing the horizontal components of the two vectors equal to the horizontal components of the four vectors.

$$0 + 60 = \left[\frac{4}{5}(\omega_{BC})^2 \times 50 \right] + \left[\frac{3}{5}(50\alpha_{BC}) \right] + \left[(\omega_{CD})^2 \times 60 \right] + [0]$$

$$60 = \left[\frac{4}{5}(3.34)^2 \times 50 \right] + \left[\frac{3}{5} \times 50\alpha_{BC} \right] + \left[(2.22)^2 \times 60 \right]$$

$$60 = 446 + 30\alpha_{BC} + 296$$

$$\alpha_{BC} = -22.7 \text{ rad/s}^2$$

The minus sign indicates that the direction of α_{BC} that we assumed was incorrect, but we can substitute this value into the next equation as a minus value.

Now consider all vertical components.

$$-500+0=\left[-\frac{3}{5}(\omega_{BC})^2\times50\right]+\left[\frac{4}{5}\times50\alpha_{BC}\right]+[0]-[60\alpha_{CD}]$$

$$-500=\left[-\frac{3}{5}(3.34)^2\times50\right]+\left[\frac{4}{5}\times50(-22.7)\right]-[60\alpha_{CD}]$$

$$-500=-335-908-60\alpha_{CD}$$

$$\alpha_{CD}=-12.4\text{ rad/s}^2\text{ }\curvearrowright$$

Therefore:

$$\alpha_{CD}=12.4\text{ rad/s}^2\text{ }\curvearrowleft$$

We can now find the linear acceleration of C by considering link CD alone (Figure 12-25).

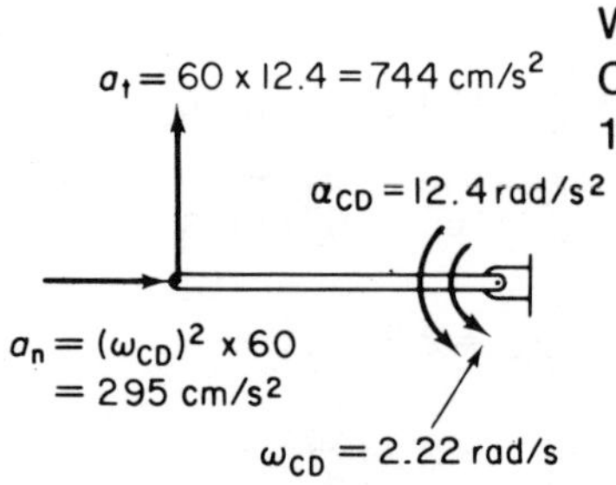

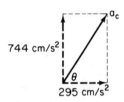

$a_t=60\times12.4=744$ cm/s^2

$\alpha_{CD}=12.4$ rad/s^2

$a_n=(\omega_{CD})^2\times60$
$=295$ cm/s^2

$\omega_{CD}=2.22$ rad/s

Figure 12-25

The total acceleration of C is (Figure 12-26):

a_C

744 cm/s^2

θ

295 cm/s^2

Figure 12-26

$$a_C=\sqrt{(295)^2+(744)^2}$$

$$=800\text{ cm/s}^2$$

$$\tan\theta=\frac{744}{295}$$

$$\theta=68.4°$$

$$a_C=8\text{ m/s}^2\text{ }\nearrow 68.4°$$

12-2   THE ROLLING WHEEL

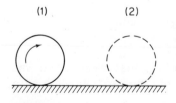

(1)   (2)

Figure 12-27

An everyday example of plane motion is a rolling wheel. If a wheel is rolled from position (1) to position (2) in Figure 12-27, it has not only rotational motion but also translational motion. If it had been pivoted at its center and were not rolling, then its motion would have been only rotational.

Since there is translational motion to the right, every point on the wheel must have some velocity to the right.

To visualize the speed at which the wheel would move to the right, consider a wheel with a radius of 0.5 m held slightly off the ground and rotated at 8 rad/s clockwise (Figure 12-28). The velocity of A is the velocity of A with respect to C, or:

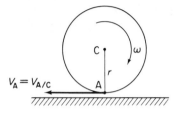

Figure 12-28

$$v_{A/C} = r\omega$$
$$= 0.5 \times 8$$
$$v_{A/C} = 4 \text{ m/s} \leftarrow$$

The wheel is now lowered to the ground while rotating. (We shall assume that there is no slippage.) Point A is stationary at this instant, but there is still a relative velocity between A and C, $v_{C/A}$.

$$v_{C/A} = -v_{A/C}$$
$$v_{C/A} = 4 \text{ m/s} \rightarrow$$

The translational speed of the center of a rolling wheel is therefore equal to $r\omega$ (Figure 12-29).

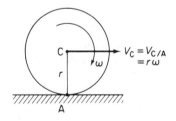

Figure 12-29

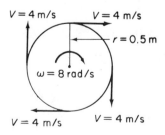

Figure 12-30

As before, plane motion can be broken into its rotational (Figure 12-30) and translational (Figure 12-31) components, and shown for given points on the wheel. Figure 12-32 shows these two motions superimposed; the result is *total plane motion.*

Note that each velocity vector is at a right angle to the line from its origin to point A. The velocity of any point on a rolling wheel

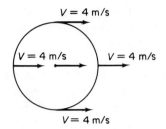

Figure 12-31

can be found by multiplying the radius distance to A by the angular speed ω. Point A is the point of contact between the wheel and the ground, and each distance measured from A is essentially a radius with a tangential velocity at a right angle to it.

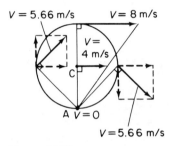

Figure 12-32

Example 12-7

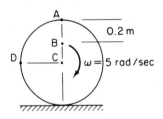

Figure 12-33

The 0.8-m diameter wheel in Figure 12-27 rolls to the right with an angular speed of 5 rad/s. Determine the velocities of points A, B, C, and D (Figure 12-33).

$$v_C = r\omega$$
$$= 0.4 \times 5$$
$$\underline{v_C = 2 \ m/s \rightarrow}$$

$$v_A = 0.8 \times 5$$
$$\underline{v_A = 4 \ m/s \rightarrow}$$
$$v_B = 0.6 \times 5$$
$$\underline{v_B = 3 \ m/s \rightarrow}$$

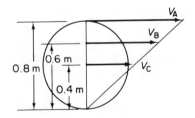

Figure 12-34

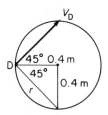

Figure 12-35

For point D:

$$r = \sqrt{(0.4)^2 + (0.4)^2}$$
$$r = 0.566 \text{ m}$$
$$v_D = 0.566 \times 5$$
$$\underline{v_D = 2.83 \text{ m/s} \nearrow 45°}$$

Example 12-8

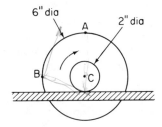

Figure 12-36

A wheel 6 in. in diameter fits through a slot and rolls on its hub, which has a diameter of 2 in. (Figure 12-36). The wheel has an angular speed of 10 rad/sec clockwise. Determine the velocities of points A, B, and C.

For point C:

$$v_C = r\omega$$
$$= 1 \times 10$$
$$\underline{v_C = 10 \text{ in./sec} \rightarrow}$$

For point A:

$$v_A = 4 \times 10$$
$$\underline{v_A = 40 \text{ in./sec} \rightarrow}$$

For point B (Figure 12-37):

$$r = \sqrt{(1)^2 + (3)^2}$$
$$r = 3.16 \text{ in}$$
$$v_B = 3.16 \times 10$$
$$\underline{v_B = 31.6 \text{ in./sec} \nearrow \frac{3}{1}}$$

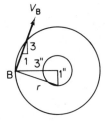

Figure 12-37

12-3   INSTANTANEOUS CENTER OF ROTATION

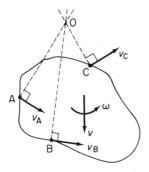

Figure 12-38

Consider a rigid body with plane motion consisting of movement downward and counterclockwise rotation. Points A, B, and C have absolute velocities as shown in Figure 12-38. This body will appear to have pure rotation if it is viewed from a point at which all the velocities are tangential velocities. Construct a radius arm perpendicular to each velocity vector. From point O and for the instant shown, the body would appear to have pure rotation with zero translational motion. The point O, about which all velocities appear as tangential velocities, is called the *instantaneous center of rotation* and has zero velocity.

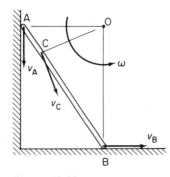

Figure 12-39

To illustrate this with a more concrete example, let us reconsider the bar in Figure 12-5 (reproduced here in Figure 12-39). For the bar to remain in contact with the wall and floor, A and B must have the velocities shown in the figure. For v_A to appear as a tangential velocity, the instantaneous center O must be somewhere on a horizontal line that passes through point A.

Similarly, the radius for tangential velocity v_B must be a vertical line at a right angle to v_B. The intersection of these two lines gives the instantaneous center at point O. For any point on the bar (including point C):

$$v = r\omega$$

or

$$\omega = \frac{v}{r}$$

$$\omega = \frac{v_A}{AO} = \frac{v_B}{OB} = \frac{v_C}{OC}$$

This ω value is also equal to the angular velocity of AB, i.e., ω_{AB}. Consider triangle AOB (Figure 12-39): all lines and points within the triangle rotate about instantaneous center O and have an angular velocity ω. Therefore:

$$\omega_{OA} = \omega_{OB} = \omega_{AB}$$

Previously, we would have solved for ω_{AB} by using:

$$v_A = v_B + v_{A/B}$$

then

$$\omega_{AB} = \frac{v_{A/B}}{AB}$$

Referring back to the rolling wheel (Section 12-2), you can see that the instantaneous center of rotation is the point at which the wheel contacts the flat surface. All velocities are tangent to radii from this instantaneous center. Note that only absolute velocities are involved and that the instantaneous center is only applicable at—as the name suggests—the instant shown.

Another word of warning: this theory does not apply to acceleration; it applies only to velocity since it is an instantaneous center of zero velocity (not necessarily zero acceleration). Do not apply it to acceleration vectors.

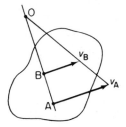

If the velocities of points A and B on a body are parallel, then there will be no point of intersection, and the result will be that similar triangles exist, such as those in Figure 12-40.

Figure 12-40

Example 12-9

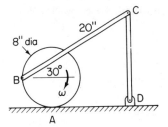

Figure 12-41

Link BC in Figure 12-41 is pinned to a cylinder rolling at 12 in./sec to the right. Calculate the velocity of pin C for the instant shown.

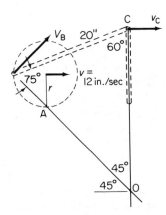

Figure 12-42

Point A is the instantaneous center of rotation for the cylinder (Figure 12-42).

$$\text{radius } AB = 5.66 \text{ in.}$$

Therefore

$$\omega \text{ about point A} = \frac{v}{r} = \frac{v_B}{AB}$$

$$= \frac{12}{4} = \frac{v_B}{5.66}$$

$$v_B = 17 \text{ in./sec}$$

We now locate point O, the instantaneous center for v_B and v_C, by drawing lines perpendicular to v_B and v_C.

$$\frac{OB}{\sin 60°} = \frac{20}{\sin 45°} = \frac{OC}{\sin 75°}$$

$$OB = 24.5 \text{ in.}$$

$$OC = 27.4 \text{ in.}$$

$$\omega \text{ about O} = \frac{v_B}{OB} = \frac{v_C}{OC}$$

$$\frac{17}{24.5} = \frac{v_C}{27.4}$$

$$\underline{v_C = 19 \text{ in./sec} \rightarrow}$$

Example 12-10

Member AC rotates at 2 rad/s counter-clockwise (Figure 12-43). Use the method of instantaneous centers to determine:

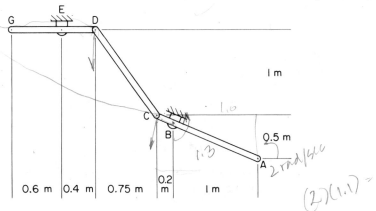

Figure 12-43

(a) velocity of point D;

(b) angular velocity of DC; and

(c) velocity of point G.

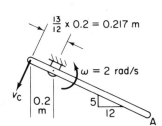

Figure 12-44

By considering member CA (Figure 12-44), we get:

$$v_C = r\omega$$

$$= 0.217 \times 2$$

$$v_C = 0.433 \text{ m/s} \quad \substack{12 \\ 5}$$

Consider member DC and locate the instantaneous center by drawing radii perpendicular to the velocities v_C and v_D (Figure 12-45).

$$\text{radius } OC = \frac{13}{5} \times 1 = 2.6 \text{ m}$$

$$\text{radius } OD = 2.4 - 0.75 = 1.65 \text{ m}$$

$$\omega = \frac{v_D}{OD} = \frac{v_C}{OC}$$

$$\frac{v_D}{1.65} = \frac{0.433}{2.6}$$

$$v_D = 0.275 \text{ m/s} \downarrow$$

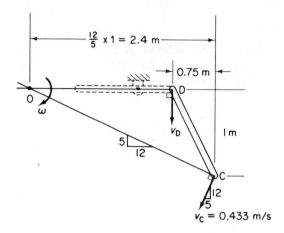

Figure 12-45

All lines in triangle ODC have the same angular velocity:

$$\omega = \frac{v_C}{OC} = \omega_{DC}$$

Therefore:

$$\omega_{DC} = \frac{0.433}{2.6}$$

$$\underline{\omega_{DC} = 0.167 \text{ rad/s } \curvearrowright}$$

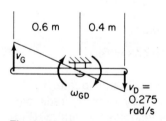

Figure 12-46

For Figure 12-46, we have:

$$\omega_{GD} = \frac{v_D}{ED} = \frac{v_G}{EG}$$

or

$$\frac{0.275}{0.4} = \frac{v_G}{0.6}$$

$$\underline{v_G = 0.412 \text{ m/s}\uparrow}$$

Example 12-11

Link AB of the mechanism shown in Figure 12-47 rotates at 5 rad/s counterclockwise. Point C, a point on link BD, is assumed to be directly over pin E for the purposes of

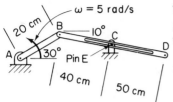

Figure 12-47

this example. Determine (a) the velocity of points B, C, and D, and (b) the angular velocity of link BD.

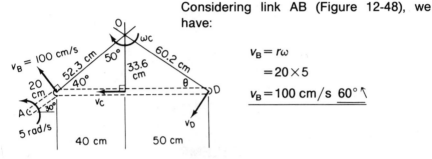

Figure 12-48

Considering link AB (Figure 12-48), we have:

$$v_B = r\omega$$
$$= 20 \times 5$$
$$\underline{v_B = 100 \text{ cm/s } 60° \nwarrow}$$

Drawing perpendicular lines from the known velocity directions of B and C, we can locate the instantaneous center of rotation O. In triangle OBC:

$$\tan 40° = \frac{OC}{40}$$
$$OC = 33.6 \text{ cm}$$
$$\cos 40° = \frac{40}{OB}$$
$$OB = 52.3 \text{ cm}$$

In triangle OCD;

$$OD = \sqrt{(33.6)^2 + (50)^2}$$
$$OD = 60.2 \text{ cm}$$

All velocities have the same angular speed ω about 0, therefore:

$$\omega_0 = \frac{v_B}{OB} = \frac{v_C}{OC} = \frac{v_D}{OD}$$
$$\frac{100}{52.3} = \frac{v_C}{33.6}$$
$$\underline{v_C = 64.3 \text{ cm/s } 10° \nwarrow}$$

and

$$\frac{100}{52.3} = \frac{v_D}{60.2}$$

$$\underline{v_D = 115 \text{ cm/s } \overline{46°}\swarrow}$$

For triangle BOD (Figure 12-48):

$$\omega_{BD} = \omega_O = \frac{v_B}{OB}$$

Therefore:

$$\omega_{BD} = \frac{100}{52.3}$$

$$\underline{\omega_{BD} = 1.91 \text{ rad/s } \curvearrowright}$$

PROBLEMS

12-1 The bucket of a front-end loader is lifted 6 ft as the tractor moves ahead 10 ft. What is the absolute or total distance moved by the bucket?

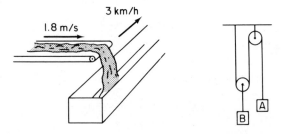

Figure P12-2   **Figure P12-3**

12-2 An automated feeding system has a conveyor that moves at 3 km/h parallel to the feed trough and has a belt speed of 1.8 m/s (Figure P12-2). Determine the absolute velocity of the material on the conveyor.

12-3 If weight A moves 4 m downward (Figure P12-3), determine the distance that A moves with respect to B.

12-4 A high-speed escalator travels at 180 ft/min. What is the absolute velocity of a man running at 700 ft/min (a) in the same direction as the escalator's motion and (b) in a direction opposite to that of the escalator's motion?

12-5 Two roller coaster cars on adjacent tracks appear on the same line of sight, as in Figure P12-5. If the velocity of car A is 4 m/s and the velocity of car B is 15 m/s, determine the velocity of car A with respect to car B.

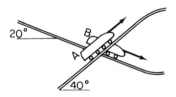

Figure P12-5

12-6 An entertainer walks forward with a velocity of 1 m/s while 3.5 m from the center of a stage that is rotating at 1.2 rpm (Figure P12-6). Determine his absolute velocity.

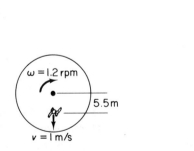

Figure P12-6

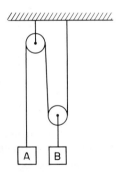

Figure P12-8

12-7 Car A, travelling north at 70 mph, meets car B, travelling south at 40 mph. Determine (a) the velocity of A with respect to B, and (b) the velocity of B with respect to A.

12-8 If the acceleration of B in Figure P12-8 is 15 ft/sec² downward, determine the acceleration of A with respect to B.

12-9 Starting from rest at the same point and at the same time, motorcycle A accelerates at 3 m/s² to the west and motorcycle B accelerates at 6 m/s² to the north. At $t=8$ sec, determine the distance, velocity, and acceleration of B with respect to A.

12-10 Starting from rest, car A accelerates at 2.5 m/s² to the east. Starting at the same point at the same time, car B accelerates at 2 m/s² to the north. At $t=10$ s, determine the distance, velocity, and acceleration of B with respect to A.

12-11 The angular velocity of AB in Figure P12-11 is 400 rpm clockwise. Determine the angular velocity of BC and the velocity of C when (a) $\theta=0°$, and (b) $\theta=90°$

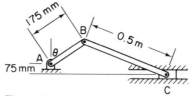

Figure P12-11

12-12 If the bicycle in Figure P12-12 is being pedalled so that it is accelerating and has a velocity of 20 mph for the instant shown, what is the absolute velocity of pedal A?

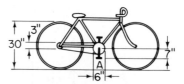

Figure P12-12

12-13 The gear drive of a food mixer is shown in Figure P12-13. The outer gear is fixed, and gear A, which has a diameter of 40 mm, rotates on the end of arm AB. Member DE is welded to gear A. If $\omega_{AB}=8$ rad/s, determine the absolute velocity of D and E at the instant shown.

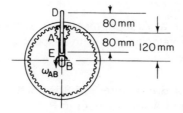

Figure P12-13

12-14 The gear in Figure P12-14 pivots at B on block A. If block A is moving to the right with a velocity of 12 mm/s, determine the velocity of C with respect to B. What is the absolute velocity of C?

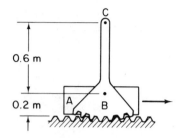

Figure P12-14

12-15 For the mechanism in Figure P12-15, AB rotates at 4 rad/s clockwise and is decelerating at 10 rad/s² clockwise. Determine the linear acceleration of C.

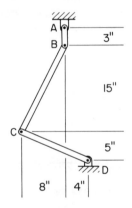

Figure P12-15

12-16 At the instant shown in Figure P12-16, AB has an angular acceleration of 8 rad/s² clockwise. Determine the acceleration of C and the angular acceleration of BC if AB has an angular velocity of 3 rad/s clockwise.

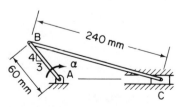

Figure P12-16

12-17   At the position shown in Figure P12-17, AB has an angular velocity of 2 rad/s counterclockwise and an angular acceleration of 6 rad/s² counterclockwise. Determine the linear acceleration of C.

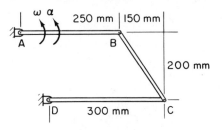

Figure P12-17

12-18   A car wheel 26 in. in diameter turns without slipping at a speed of 900 rpm. What is the speed of the car?

12-19   The cylinder shown in Figure P12-19 rolls to the right with a velocity of 6 m/s. For the instant shown, determine (a) the angular velocity of the cylinder, and (b) the linear velocity of point B.

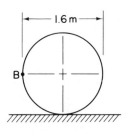

Figure P12-19

12-20   A cylinder 2 m in diameter rolls to the right on a horizontal surface with a velocity of 0.6 m/s. For the instant shown in Figure P12-20, determine (a) the angular velocity of the cylinder, and (b) the linear velocity of point B.

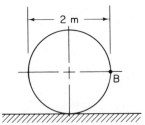

Figure P12-20

12-21 If cylinder A in Figure P12-21 rolls to the left at 20 ft/sec, determine the velocity of point B at the position shown.

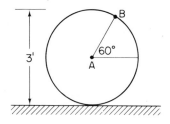

Figure P12-21

12-22 A cord is wound in the slot of cylinder A in Figure P12-22. Mass B moves downward with a velocity of 6 m/s. Assume no slipping of the cylinder and determine the velocities of points D, E, and C on cylinder A. If B drops 4 m, how far does cylinder A move to the right?

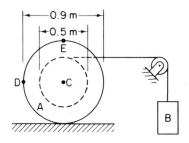

Figure P12-22

12-23 Wheel A in Figure P12-23 rolls without slipping; weight B has a velocity of 20 ft/sec downward. Determine the velocities of points D, E, and C on cylinder A. If B drops 10 ft, how far does cylinder A move to the right?

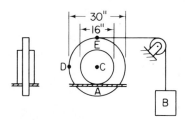

Figure P12-23

12-24 Cylinder A in Figure P12-24 rolls 5 m down the slope. What distance is mass B lifted?

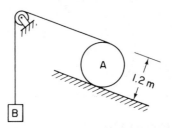

Figure P12-24

12-25 In order to pull and winch a car from a ditch, a cable is wound around the hub of a tractor rear wheel as is shown in Figure P12-25. Assume no slippage of the tractor wheel and determine the distance that the tractor must travel to pull the car 20 m.

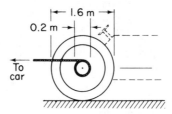

Figure P12-25

12-26 Starting from rest, weight B in Figure P12-26 drops 16 in. in 4 sec. Determine the angular velocity and the angular acceleration of the cylinder at $t = 4$ sec.

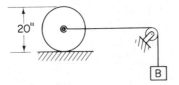

Figure P12-26

12-27 A barrel 30 in. in diameter rolls down a slope with an acceleration of 5 ft/sec². Determine the barrel's angular acceleration.

12-28 If the wheel in Figure P12-28 rolls to the left with an angular velocity of 6 rad/s, determine (a) the velocity of B, and (b) the angular velocity of AB.

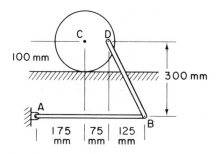

Figure P12-28

12-29 Determine the angular velocity of the roller in Figure P12-29 if the velocity of B is 10 m/s up the slope.

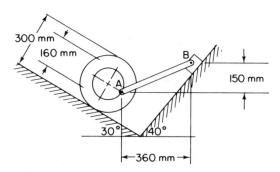

Figure P12-29

12-30 The wheel shown in Figure P12-30 rolls to the right at 40 in./sec. Determine the angular velocities of AB and BC.

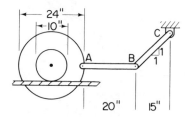

Figure P12-30

12-31   By the method of instantaneous centers, determine the velocities of points A, B, and D on the wheel shown in Figure P12-31. The velocity of C is 10 m/s to the right.

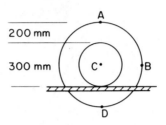

Figure P12-31

12-32   Pin C of the linkage in Figure P12-32 has a velocity of 5 m/s downward. Use the method of instantaneous centers to determine the velocity of B and the angular velocity of AB.

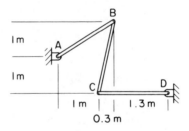

Figure P12-32

12-33   Slider B in Figure P12-33 has a velocity of 3 ft/min upward. By the method of instantaneous centers, determine the velocity of point A.

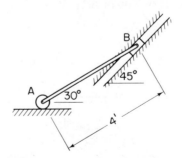

Figure P12-33

12-34 If slider C in Figure P12-34 moves downward at 0.7 m/s, determine (a) the angular velocity of AB, and (b) the velocity of D.

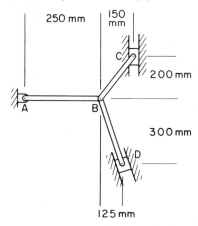

Figure P12-34

12-35 Block B in Figure P12-35 has a velocity of 10 ft/sec to the left. Determine the velocities of A and C.

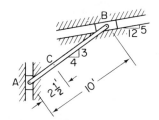

Figure P12-35

12-36 Link BC in Figure P12-36 is pinned to a cylinder rolling at 0.8 m/s to the left. Determine the angular velocity of BC and the linear velocity of pin C.

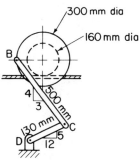

Figure P12-36

12-37 Roller A in Figure P12-37 has a velocity of 2 m/s to the right. Use the method of instantaneous centers to determine the velocity of point D.

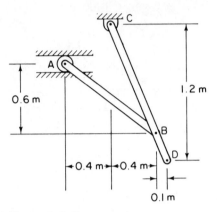

Figure P12-37

12-38 The wheel in Figure P12-38 turns at 8 rad/sec clockwise. Determine the velocity of B.

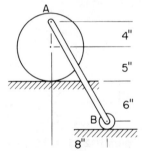

Figure P12-38

12-39 Wheel A in Figure P12-39 rolls to the right at 120 mm/s. Determine the velocity of C.

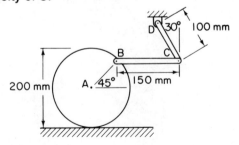

Figure P12-39

12-40 Member ED in Figure P12-40 rotates at 2 rad/sec clockwise. Determine the velocities of points B and C.

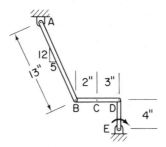

Figure P12-40

12-41 The angular velocity of AB in Figure P12-41 is 4 rad/s clockwise. Use instantaneous centers to determine the velocity of points C and E.

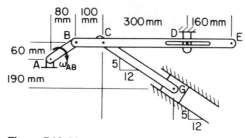

Figure P12-41

12-42 Use instantaneous centers to determine the velocity of point G if $\omega_{AB} = 4$ rad/s ↺ (Figure P12-41).

12-43 Use the method of instantaneous centers to solve Problem 12-20.

12-44 Use the method of instantaneous centers to solve Problem 12-21.

12-45 Use the method of instantaneous centers to solve Problem 12-28.

12-46 Use the method of instantaneous centers to solve Problem 12-29.

13

KINETICS

Kinetics concerns not only velocity and acceleration but also the accompanying unbalanced forces that cause the motion. At this point, it would be convenient to summarize Newton's three laws of motion.

1. Every object remains at rest or moves with constant velocity in a straight line unless an unbalanced force acts upon it.

This is the law that we applied in statics when we balanced force systems. Another case of the same force balance that we have not yet considered is that of an object travelling at constant velocity in a straight line.

2. A body that has a resultant unbalanced force acting upon it behaves as follows:
    (a) the acceleration is proportional to the resultant force;
    (b) the acceleration is in the direction of the resultant force; and
    (c) the acceleration is inversely proportional to the mass of the body.

3. For every action there is an equal and opposite reaction.

Kinetics, the study of unbalanced forces causing motion, can be analyzed by three methods (as the following chapters will show):

1. Inertia force or torque (dynamic equilibrium)
2. Work and energy
3. Impulse and momentum

This chapter will cover the first method as applied to both *linear* and *angular motion*. (Linear motion is another term for rectilinear or translational motion.)

13-1   LINEAR INERTIA FORCE

Newton's second law can be examined more closely and placed in equation form.

Let:

a = acceleration
F = resultant force
m = mass of body
W = force of gravity (weight in the English system)

1. $a \propto F$ or $a = F \times$ constant
2. $a \rightarrow$ and $F \rightarrow$
3. $a \propto \dfrac{1}{m}$ or $a = \dfrac{\text{constant}}{m}$

These three statements can be combined and stated in one equation:

$$F = ma \tag{13-1}$$

where in SI system

F = force in newtons (N)
m = mass in kilograms (kg)
a = acceleration in m/s^2

This is consistent with the original definition of 1 newton being the force that causes a mass of 1 kg to accelerate at 1 m/s^2.

$$1 \text{ N} = 1 \text{ kg} \times 1 \text{ m/s}^2$$
$$1 \text{ N} = 1 \text{ kg} \cdot \text{m/s}^2$$

This force can also be the force of gravity on an object (customarily called weight). Taking the acceleration of gravity as 9.81 m/s^2 and using W to represent weight or force of gravity, from

$$\text{force} = \text{mass} \times \text{acceleration}$$

we get:

$$\text{weight} = \text{mass} \times \text{acceleration of gravity}$$
$$W = m \times g$$

For a 1 kg mass:

$$W = 1 \text{ kg} \times 9.81 \text{ m/s}^2$$
$$W = 9.81 \text{ N}$$

Although mass and weight are often confused in the English system, they are distinguished as follows.

The *weight* of an object is a measure of the pull of gravity on it. The acceleration due to gravity, g, may vary depending on the location of the object on the earth's surface. Since the weight of an object can vary with gravity, another term, *mass*, a quantity that does not change with changing gravity, is used to describe an object. Mass is constant for an object and is defined by the following equation:

$$\text{mass} = \frac{\text{weight}}{\text{acceleration due to gravity}}$$

$$m = \frac{W}{g} \tag{13-2}$$

where

$$m = \frac{\text{lb}}{\text{ft/sec}^2} = \text{slugs}$$

$$W = \text{lb}$$

$$g = 32.2 \text{ ft/sec}^2 \quad \text{(an approximate value for our calculations)}$$

From Equation 13-2, you can see that, although there may be one-half the weight at a particular location due to one-half the acceleration or pull of gravity, the mass would still be the same.

In summary, we now have a situation in which a force or unbalance of several forces causes a body to move with changing velocity; that is, the body accelerates. The equations describing the body, the force, and the acceleration are:

$$F = ma, \quad \text{for a single force}$$

or

$$\Sigma F = ma, \quad \text{for several forces}$$

13-2 LINEAR INERTIA FORCE—DYNAMIC EQUILIBRIUM

There are two methods of dealing with inertia and acceleration for linear motion. The *inertia-force method of dynamic equilibrium* will be used here since it often gives rise to a shorter, easier solution. The other method, which will not be covered, involves summing all forces to obtain the force that causes unbalance and equating that sum with mass times acceleration.

In order to visualize the inertia-force method, recall Newton's third law: For every action there is an equal and opposite reaction.

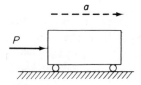

Figure 13-1

Suppose that we have action or an acting force P that causes acceleration of the block in Figure 13-1 to the right. Where is the opposite reaction or opposing force? The opposing force is the inertia force and is equal to ma. Keep in mind that inertia of an object tends to maintain the present state of the object; it opposes change. If the object is stationary, inertia opposes motion. If the object is moving at constant velocity or in a straight line, inertia opposes velocity or direction change. With any of these types of change, we have acceleration and the resulting inertia force. The direction in which the inertia force acts is always opposite to the direction of the acceleration.

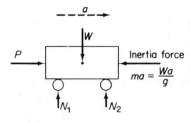

Figure 13-2

The unbalanced force P is opposed by the inertia force, which is equal to mass times acceleration and is opposite in direction to the acceleration (Figure 13-2). The other vectors of weight and normal forces are shown as they would be in statics. The block is now in dynamic equilibrium, and it too can be treated exactly as it was in statics. All methods and equations of statics apply.

Just as the weight of an object is taken as acting through the center of gravity, the inertia force also is thought of as acting through the center of gravity or mass center.

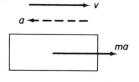

Figure 13-3

To illustrate that the inertia force always acts in a direction opposite to acceleration, consider a block moving to the right but decelerating (Figure 13-3). Deceleration is minus acceleration, but rather than showing − a to the right, we show + a to the left. The inertia force tends to keep the block moving; it is opposite to acceleration and is therefore shown acting to the right.

Just remember that all forces must be shown on the free-body diagram just as in statics. By adding an additional force to account for inertia, we produce a situation of dynamic equilibrium, and all the previous methods apply.

Example 13-1

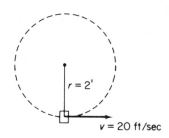

Figure 13-4

A 5-lb weight is swung in a vertical circle on the end of a 2-ft rope. If the velocity of the weight at the bottom of the circle is 20 ft/sec, determine the tension in the rope at this point (Figure 13-4).

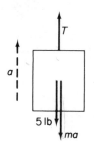

Figure 13-5

The free-body diagram of the weight is drawn, just as in statics but with the addition of the inertia force. This gives us dynamic equilibrium (Figure 13-5).

The acceleration in this case is normal acceleration.

$$a_n = \frac{v^2}{r} = \frac{400}{2} = 200\,\text{ft/sec}^2$$

$$\Sigma F_y = 0$$

$$T - 5 - \left(\frac{5}{32.2} \times 200\right) = 0$$

$$\underline{T = 36\ \text{lb}}$$

Example 13-2

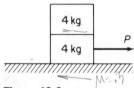

Figure 13-6

Two 4-kg mass blocks are placed one on top of the other. The coefficient of static friction for all surfaces is 0.3. Calculate the minimum P required to pull the bottom block out from beneath the top one without moving the top one horizontally (Figure 13-6).

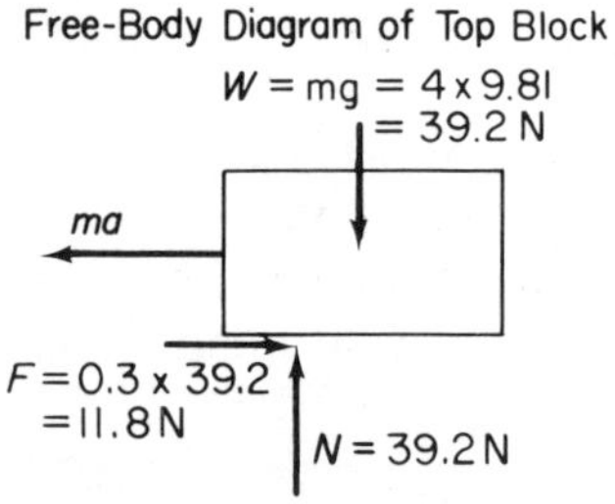

Free-Body Diagram of Top Block

$$W = mg = 4 \times 9.81$$
$$= 39.2\ \text{N}$$

ma

$$F = 0.3 \times 39.2$$
$$= 11.8\,\text{N}$$

$N = 39.2\,\text{N}$

Figure 13-7

The free-body diagram of the top block (Figure 13-7) shows that the inertia must be sufficient to overcome the friction force of the bottom block on the top one.

$$\Sigma F_x = 0$$

$$0 = 11.8 - ma$$
$$a = \frac{11.8}{m}$$
$$= \frac{11.8}{4}$$
$$a = 2.95\,\text{m/s}^2$$

Free-Body Diagram of Bottom Block

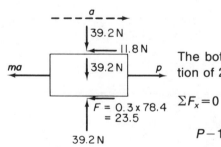

39.2 N
11.8 N
ma
39.2 N P
$$F = 0.3 \times 78.4$$
$$= 23.5$$
39.2 N

Figure 13-8

The bottom block must have an acceleration of 2.95 m/s² or greater.

$$\Sigma F_x = 0 \qquad\qquad \text{(Figure 13-8)}$$

$$P - 11.8 - 23.5 - (4 \times 2.95) = 0$$
$$\underline{P = 47.1\,\text{N} \rightarrow}$$

Example 13-3

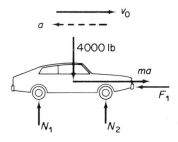

Figure 13-9

A car bumper is designed to bring a 4000-lb car to a stop from a speed of 5 mph while deforming 6 in. Assume constant deceleration and determine the average force on the bumper during this stop (Figure 13-9).

$$v_o = \frac{5}{60} \times 88 = 7.33 \,\text{ft/sec}$$

$$v^2 = v_o^2 + 2as$$

$$0 = (7.33)^2 + 2a \times 0.5$$

$$a = -53.8 \,\text{ft/sec}^2 \leftarrow$$

$$a = 53.8 \,\text{ft/sec}^2 \leftarrow$$

$$\Sigma F_x = 0 \qquad\qquad\qquad \text{(Figure 13-9)}$$

$$ma - F_1 = 0$$

$$F_1 = \frac{4000}{32.2} \times 53.8$$

$$\underline{F_1 = 6680 \,\text{lb}}$$

The bumper must withstand a force of 6680 lb.

Example 13-4

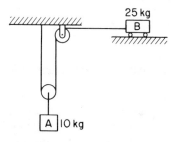

Figure 13-10

Neglect the inertia of the pulleys and the rolling resistance of mass B and determine the acceleration of masses A and B when the system is relea∪ed from rest (Figure 13-10).

Notice that, if mass A drops 1 m, mass B will move 2 m: B moves twice as far as A does in the same time interval; thus, the acceleration of B is twice that of A.

$$a_B = 2a_A$$

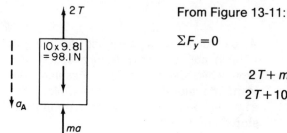

Figure 13-11

From Figure 13-11:

$$\Sigma F_y = 0$$

$$2T + ma_A - 98.1 = 0$$
$$2T + 10a_A - 98.1 = 0$$
$$T = 49 - 5a_A \quad (1)$$

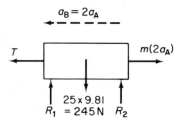

Figure 13-12

In Figure 13-12, there is twice the acceleration; therefore, the mass times acceleration term becomes $m(2a)$.

$$\Sigma F_x = 0$$

$$m(2a_A) = T$$
$$25 \times 2a_A = T$$
$$T = 50a_A \quad (2)$$

Equating Equations (1) and (2), we get:

$$50a_A = 49 - 5a_A$$
$$a_A = \frac{49}{55} = 0.89 \text{ m/s}^2$$

The acceleration of A is 0.89 m/s² downward, and the acceleration of B is 1.78 m/s² to the left.

13-3 ANGULAR INERTIA

The main limitation that we impose on the angular inertia of a rotating body in this section is that the axis of rotation coincide with the center of mass. We therefore have a homogeneous body with an angular motion and inertia about the body's center of mass. This by no means covers all types of rotation. Section 13-5 will deal with the topic in more depth.

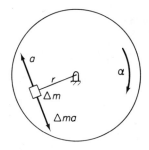

Figure 13-13

Consider the portion of the body (Δm) shown in Figure 13-13; the larger Δm is and the larger the radius, the larger the body's angular inertia will be. To give this wheel a clockwise angular acceleration, a clockwise torque would have to be applied. This torque would accelerate many particles of mass Δm, each at its own radius r. Each Δm has an acceleration tangent to the radius and an inertia force (Δma) in the opposite direction. The torque for each Δm is (Δma)r where

$$a = r\alpha \quad \text{or} \quad \text{torque} = (\Delta mr\alpha)r$$

The total torque that will accelerate all elements of Δm is:

$$\text{torque} = \Sigma(\Delta mr\alpha)r$$
$$\text{torque} = \Sigma r^2 \alpha \Delta m$$

Since α is constant:

$$\text{torque } t = \alpha\Sigma r^2 \Delta m$$

From Section 9-5, (Mass Moment of Inertia), we have:

$$I = \Sigma r^2 \Delta m$$

therefore:

$$\text{torque } t = I_c \alpha \qquad\qquad (13\text{-}3)$$

where

> torque is in N·m
> I_c = mass moment of inertia about the center of mass in kg·m^2
> α = angular acceleration in rad/s^2

In the English system:
torque is in lb-ft
I_c = mass moment of inertia about the center of mass in slug-ft^2 or ft-lb-sec^2
α = angular acceleration in rad/sec^2

I can be obtained from Table 9-2 or from the radius of gyration equation (Section 9-7).

$$k = \sqrt{\frac{I}{m}} \quad \text{or} \quad I = k^2 m$$

Example 13-5

What torque is required to accelerate a wheel about its center ($I_C = 2$ ft-lb-sec^2) to an angular acceleration of 30 rad/sec^2?

$$\text{torque } t = I_C \alpha$$
$$= 2 \times 30$$
$$\underline{\text{torque } t = 60 \text{ lb-ft}}$$

Example 13-6

A rotor with a mass moment of inertia (I_C) of 6 kg·m^2 about its center of mass has a torque of 90 N·m applied to it. Determine the angular acceleration of the rotor.

$$\text{torque } t = I_C \alpha$$
$$90 = 6\alpha$$
$$\underline{\alpha = 15 \text{ rad/s}^2}$$

13-4   ANGULAR DYNAMIC EQUILIBRIUM

With angular dynamic equilibrium, we are again faced with a situation analogous to rectilinear and curvilinear motion (Section 13-2). We have shown an inertia force opposite in direction to acceleration. This gave us dynamic equilibrium and the ability to solve in the same manner as previous statics problems.

The same situation exists here since we now show the angular inertia torque opposite in direction to the angular acceleration; thus dynamic equilibrium results. But there is an important distinction: whereas we had a *force* (*ma*), for rectilinear and curvilinear motion, we have a *torque* (*Iα*) for angular motion. The importance of this distinction will become more evident after you have dealt with the following examples.

Example 13-7

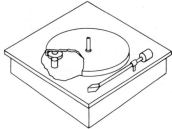

Figure 13-14

The turntable of the record changer shown in Figure 13-14 weighs 8 lb and is accelerated from rest to a speed of 33 1/3 rpm in 3 sec. It is driven by an electric motor with a rubber drive wheel 2 in. in diameter pressed against the inside of the turntable rim which is 11 in. in diameter. Assume negligible inertia torque of the rubber wheel and motor and determine the torque that the motor must supply to provide the acceleration described.

Free-Body Diagram
of Turntable

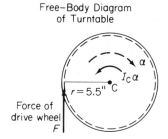

Force of
drive wheel
F

Figure 13-15

Assume that the turntable is a flat circular plate; therefore (from Table 9-2):

$$I_C = \frac{1}{2} mr^2$$

$$I_C = \frac{1}{2} \times \frac{8}{32.2} \times \left(\frac{5.5}{12}\right)^2$$

$$= 0.0261 \text{ ft-lb-sec}^2$$

$$\omega = \omega_o + \alpha t$$

$$\frac{33.3 \times 2\pi}{60} = 0 + \alpha \times 3$$

$$\alpha = 1.16 \text{ rad/sec}^2$$

$$\Sigma M_C = 0 \qquad\qquad\qquad \text{(Figure 13-15)}$$

$$-Fr + I_C \alpha = 0$$

$$-\left(F \times \frac{5.5}{12}\right) + (0.0261 \times 1.16) = 0$$

$$F = 0.066 \text{ lb}$$

This force is supplied by a wheel 2 in. in diameter; therefore, the motor torque is:

$$F \times r = 0.066 \times 1$$

$$\underline{F \times r = 0.066 \text{ lb-in.}}$$

Example 13-8

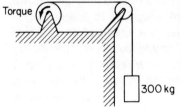

Figure 13-16

A power-driven winch is used to raise a mass of 300 kg with an acceleration 2 m/s². The winch drum is 0.5 m in diameter and has a mass moment of inertia about its center I_C equal to 8 kg·m². What torque must be applied to the winch drum (Figure 13-16)?

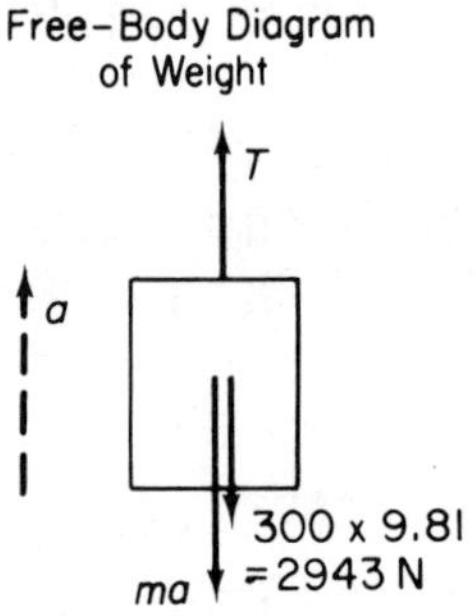

Free-Body Diagram
of Weight

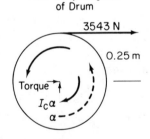

Figure 13-17

The rope tension T must first be found. Use a free-body diagram of the weight (Figure 13-17).

$$\Sigma F_y = 0$$

$$T - 2943 - (300 \times 2) = 0$$
$$T = 3543 \text{ lb}$$

Free-Body Diagram
of Drum

3543 N

0.25 m

Torque

$I_C \alpha$

α

Figure 13-18

For the drum:

$$a = r\alpha$$
$$\alpha = \frac{a}{r} = \frac{2}{0.25} = 8 \text{ rad/s}^2 \text{ ↻}$$

Taking moments about the center of the drum, we get:

$$\Sigma M_C = 0$$

$$\text{torque} - (3543 \times 0.25) - (8 \times 8) = 0$$
$$\text{torque} = 950 \text{ N·m ↻}$$

Note that $I_C \alpha$ is opposite in direction to α. Also note that $I_C \alpha$ is a torque and is not multiplied by a radius as is the force of 3543 N.

Example 13-9

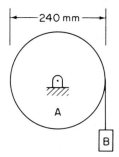

Figure 13-19

Wheel A with a mass of 22 kg has a radius of gyration of 180 mm. Mass B has a mass of 10 kg. If the system starts from rest and has no bearing friction, determine the angular acceleration of A and the tension in the rope (Figure 13-19).

Your training in statics might lead to two possible pitfalls here:

1. The tension in the rope is not $10 \times 9.81 = 98.1$ N; rather, it is somewhat less because B is accelerating downward.
2. The acceleration of B is not 9.81 m/s²; it is less due to its being restrained by the rope, which in turn is accelerating cylinder A.

Free–Body Diagram of B

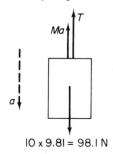

$10 \times 9.81 = 98.1$ N

Figure 13-20

For the weight B (Figure 13-20), we have:

$$\Sigma F_y = 0$$

$$T + 10a - 98.1 = 0$$
$$T + 10a = 98.1$$

but

$$a = r\alpha = 0.12\alpha$$
$$T + 1.2\alpha = 98.1 \qquad (1)$$

Free–Body Diagram of A

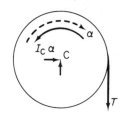

Figure 13-21

Consider the cylinder now (Figure 13-21).

$$I_c = k^2 m \qquad \text{(Eq. 9-7)}$$
$$= (0.18)^2 \times 22$$
$$I_c = 0.713 \text{ kg} \cdot \text{m}^2$$

$$\Sigma M_c = 0$$

$$-(T \times 0.12) + 0.713\alpha = 0$$
$$T = 5.94\alpha \qquad (2)$$

Substituting Equation 2 into Equation 1, we get:

$$5.94\alpha + 1.2\alpha = 98.1$$
$$\underline{\alpha = 13.7 \text{ rad/s}^2 \,\circlearrowleft}$$

and

$$\underline{T = 81.7 \text{ N}}$$

Example 13-10

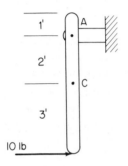

Figure 13-22

A 6-ft slender rod weighing 64.4 lb is initially at rest when the force of 10 lb is applied as shown in Figure 13-22. Determine the reactions at A and the angular acceleration of the rod for this instant.

The interia of the rod will resist motion, so its mass moment of inertia must be found. From Table 9-2, the mass moment of inertia about the center of the rod is:

$$I_c = \frac{1}{12} ml^2$$

$$= \frac{1}{12} \times \frac{64.4}{32.2} \times (6)^2$$

$$I_c = 6 \text{ ft-lb-sec}^2$$

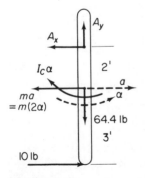

Figure 13-23

Show the rod in dynamic equilibrium (Figure 13-23).

$$a = r\alpha = 2\alpha$$

$$\Sigma F_y = 0$$

$$A_y = 64.4 \text{ lb}$$

$$\Sigma M_A = 0$$

$$(10 \times 5) - 6\alpha - \left(\frac{64.4}{32.2} \times 2\alpha\right) = 0$$

$$\underline{\alpha = 3.57 \text{ rad/s} \circlearrowleft}$$

$$\Sigma F_x = 0$$

$$A_x + \frac{64.4}{32.2}(2 \times 3.57) - 10 = 0$$

$$= -4.29 \text{ lb} \leftarrow$$

$$\underline{A_x = 4.29 \text{ lb} \rightarrow}$$

13-5 PLANE MOTION

There are many types of plane motion. The rectilinear, curvilinear, and angular motions that we have been considering separately are basically isolated or restricted forms of plane motion. Although we will be unable to cover all types of plane motion, the following list will illustrate to you the scope of this class of motion and the portion of it that we are considering.

(A) Constrained plane motion

    1. Translational

        (a) Rectilinear
        (b) Curvilinear

    2. Centroidal rotation (angular)—the axis of rotation coincides with the mass center.

    3. Translational and centroidal rotation—a rolling wheel or connecting rod are typical of this motion. There is a definite relationship between the translational acceleration and the angular acceleration.

    4. Translational and noncentroidal rotation

(B) Unconstrained plane motion

    1. Translational and unrelated centroidal rotation—a wheel simultaneously rolling and sliding is typical of this motion. There is no relationship between the translational acceleration and the angular acceleration.

As you may already have noticed, the kinetics of two of the motions listed in the above outline, translational motion (A1) and centroidal rotation (A2), were covered in Sections 13-1 to 13-4 inclusive. These forms of motion are combined to give the plane motion of (A3), translational and centroidal rotation.

With respect to kinetics, the main features of this latter type of plane motion are as follows:

1. There is a definite relationship between the angular acceleration and the translational acceleration (a_x or a_y).
2. It can be resolved into its component motions, translation and rotation.
3. Using free-body diagrams and the principle of dynamic equilibrium, we can use the three equations below:

$$\Sigma F_x = 0$$
$$\Sigma F_y = 0$$
$$\Sigma M = 0$$

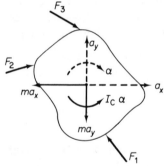

Moments can be taken about any point although the instantaneous center of rotation is often used. Figure 13-24 shows an unbalance of forces causing three simultaneous accelerations of a body.

Figure 13-24

Example 13-11

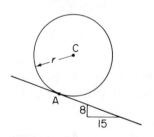

A solid cylinder 4 ft in diameter weighing 96.6 lb rolls down the slope shown in Figure 13-25 without slipping. Determine the angular acceleration.

Figure 13-25

From Table 9-2, the mass moment of inertia of a cylinder about its center of mass C is found by the following equation.

$$I_C = \frac{1}{2} mr^2$$

$$= \frac{1}{2} \times \frac{96.6}{32.2} \times (2)^2$$

$$I_C = 6 \text{ ft-lb-sec}^2$$

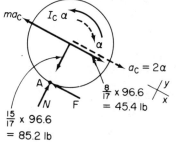

Figure 13-26

Dynamic equilibrium is shown in Figure 13-26; there, you can see that both translational inertia (ma_C) and angular inertia ($I_C\alpha$) are involved.

Since there is no slippage:

$$a_C = r\alpha = \frac{24}{12}\alpha = 2\alpha$$

Note that, since the coefficient of friction and the friction force are unknown, we cannot use either.

$$\Sigma F_x = 0$$

or

$$\Sigma M_C = 0$$

Moments are taken about A, and the torque $I_C\alpha$ is treated as a couple.

$$\Sigma M_A = 0$$

$$+ I_C\alpha + 2(ma_C) - 45.4 \times 2 = 0$$

$$6\alpha + \frac{96.6}{32.2} \times 2\alpha \times 2 = 90.8$$

$$\underline{\underline{\alpha = 5.05 \text{ rad/sec}^2 \; \curvearrowright}}$$

Example 13-12

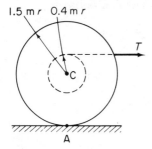

Figure 13-27

A 20-kg cylinder has a cord wound in a 0.8 m diameter groove at its center (Figure 13-27). Assume that there is no slippage and that a tension of 100 N is applied. Determine (a) the acceleration a_C of the center of mass of the wheel, and (b) the minimum coefficient of static friction.

From Table 9-2, we have:

$$I_C = \frac{1}{2}mr^2$$

$$= \frac{1}{2} \times 20 \times (1.5)^2$$

$$I_C = 22.5 \text{ kg} \cdot \text{m}^2$$

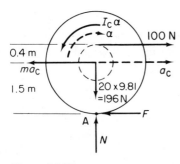

Figure 13-28

As shown in Figure 13-28:

$$a_C = r\alpha$$

$$\alpha = \frac{a_C}{1.5}$$

Taking moments about A, we get:

$$\Sigma M_A = 0$$

$$ma_C \times 1.5 + I_C\alpha - (100 \times 1.9) = 0$$

$$(20a_C \times 1.5) + \left(22.5 \times \frac{a_C}{1.5}\right) = 100 \times 1.9$$

$$\underline{a_C = 4.22 \text{ m/s}^2 \rightarrow}$$

Now solving for friction, we obtain:

$$\Sigma F_x = 0$$

$$100 - ma_C - F = 0$$

$$F = 100 - (20 \times 4.22)$$

$$F = 15.6 \text{ N} \leftarrow$$

$$\Sigma F_y = 0$$

$$N = 196 \text{ N}$$

$$\mu = \frac{F}{N} = \frac{15.6}{196} = 0.079$$

Example 13-13

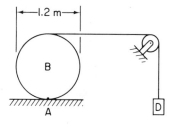

Figure 13-29

Cylinder B has a mass of 40 kg and is 1.2 m in diameter. Block D has a mass of 30 kg (Figure 13-29). Assume the mass and friction of the cable and pulley to be negligible. If this system is released from rest, determine the tension in the cable and the acceleration of D. (There is no slippage at A.)

Considering the cylinder first, we have:

$$I_C = \frac{1}{2} mr^2 = \frac{1}{2} \times 40 \times (0.6)^2$$

$$I_C = 7.2 \text{ kg} \cdot \text{m}^2$$

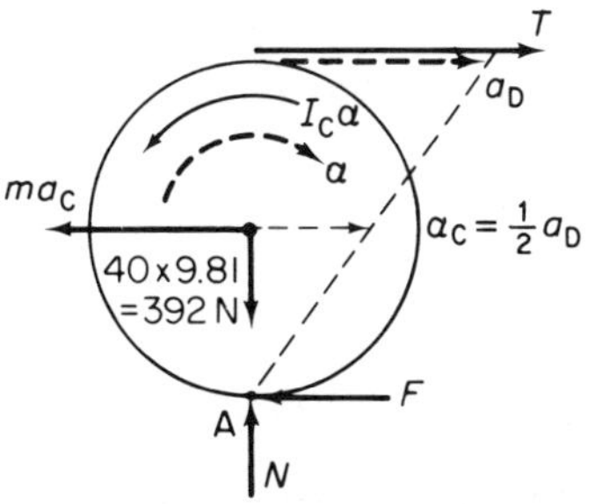

Show both the angular and the linear acceleration with their corresponding inertia torque and inertia force (Figure 13-30).

Note that the point at the top of the cylinder at which T is acting has twice the acceleration of a_C since A is the instantaneous center of rotation. This is the same as the acceleration of D or:

$$a_D = 2 a_C$$

$$a_C = \frac{1}{2} a_D$$

and also

$$a_C = r\alpha$$

$$\alpha = \frac{a_C}{r} = \frac{a_D}{2r}$$

Taking moments about A will give an equation with the two required unknowns, a_D and T.

Figure 13-30

$$\Sigma M_A = 0$$

$$(-T \times 1.2) + (ma_C \times 0.6) + I_C \alpha = 0$$

$$1.2T = \left(40 \times \frac{1}{2} a_D \times 0.6\right) + \left(7.2 \times \frac{a_D}{2 \times 0.6}\right)$$

$$T = 15 a_D \qquad (1)$$

Drawing a free-body diagram of D (Figure 13-31), we have:

$$\Sigma F_y = 0$$

$$T + ma_D - 294 = 0$$

$$T + 30 a_D = 294$$

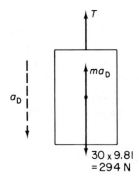

30 x 9.81
= 294 N

Figure 13-31

Substituting Equation (1), we have:

$$15 a_D + 30 a_D = 294$$

$$\underline{a_D = 6.54 \text{ m/s}^2}$$

and

$$T = 15 \times 6.54$$

$$\underline{T = 98.1 \text{ N}}$$

Example 13-14

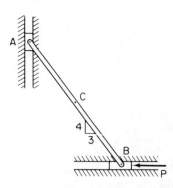

Figure 13-32

The rod AB in Figure 13-32 is 10 in. long and weighs 16.1 lb. For the instant shown, A is at rest and has an acceleration of 15 ft/sec² downward. Neglect the weight and friction of the slider blocks at A and B. Determine (a) the force P that is partially restraining the motion, and (b) the reaction forces at A and B.

The first items to be solved for will be the vertical and horizontal accelerations of the rod's center of mass, C. We must determine the angular acceleration as well.

The relative acceleration can be written (Figure 13-33).

$$a_A = a_B + a_{A/B}$$

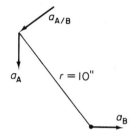

Figure 13-33

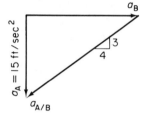

Figure 13-34

From the vector triangle of Figure 13-34, we have:

$$a_B = \frac{4}{3} \times a_A = \frac{4}{3} \times 15 = 20 \text{ ft/sec}^2$$

$$a_{A/B} = 25 \text{ ft/sec}^2$$

Since

$$a_{A/B} = r\alpha$$

$$\alpha = \frac{25}{10/12} = 30 \text{ rad/sec}^2$$

Point C has one-half the horizontal and vertical acceleration that the ends have; therefore

$$a_x = 10 \text{ ft/sec}^2$$

$$a_y = 7.5 \text{ ft/sec}^2$$

From Table 9-2, I_C for the rod is:

$$I_C = \frac{1}{12} ml^2$$

$$= \frac{16.1}{12 \times 32.2} \left(\frac{10}{12}\right)^2$$

$$I_C = 0.0289 \text{ ft-lb-sec}^2$$

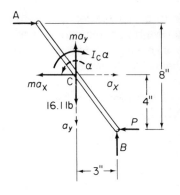

Figure 13-35

Now turn to Figure 13-35, which shows the dynamic equilibrium of the rod. We can apply any one of the three main equations:

$$\Sigma F_y = 0; \quad \Sigma F_x = 0; \quad \text{or} \quad \Sigma M = 0$$

$$\Sigma F_y = 0$$

$$B + ma_y - 16.1 = 0$$

$$B + \left(\frac{16.1}{32.2} \times 7.5\right) - 16.1 = 0$$

$$\underline{B = 12.35 \text{ lb}\uparrow}$$

$\Sigma M_B = 0$

$$-\left(A \times \frac{8}{12}\right) - I_c\alpha - \left(ma_y \times \frac{3}{12}\right) + \left(ma_x \times \frac{4}{12}\right) + \left(16.1 \times \frac{3}{12}\right) = 0$$

$$-\left(A \times \frac{8}{12}\right) - (0.0289 \times 30) - \left(\frac{16.1}{32.2} \times 7.5 \times \frac{3}{12}\right)$$

$$+ \left(\frac{16.1}{32.2} \times 10 \times \frac{4}{12}\right) + \left(16.1 \times \frac{3}{12}\right) = 0$$

$$\underline{A = 5.84 \text{ lb}\rightarrow}$$

$\Sigma F_x = 0$

$$5.84 - P - \left(\frac{16.1}{32.2} \times 10\right) = 0$$

$$\underline{P = 0.84 \text{ lb}\leftarrow}$$

The main restriction placed on the above example was that rod AB was at rest. In Figure 13-33, there was one relative acceleration, $a_{A/B}$, at point A and it was perpendicular to AB. If AB had had an angular velocity, then there would have been a normal component of $a_{A/B}$ acting along AB.

PROBLEMS

13-1 Determine the acceleration of the 150-lb block in Figure P13-1 if the coefficient of kinetic friction is 0 4.

Figure P13-1

13-2 A 50-kg block is accelerated horizontally from rest by a 600-N horizontal force. If the coefficient of friction between the block and the horizontal surface is 0.2, determine the acceleration of the block.

13-3 A 6-kg mass is whirled in a vertical circle on the end of a 1.2 m rope. What is the maximum and minimum tension in the rope during one revolution if it has a constant angular velocity of 5 rad/s?

13-4 Determine the acceleration of block A down the slope in Figure P13-4.

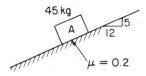

Figure P13-4

13-5 What force does a 180-lb man exert on the floor of an elevator that is moving downward and decelerating at 15 ft/sec²?

13-6 Determine the downward acceleration of an elevator so that a mass of 50 kg exerts only a force of 400 N upon its floor.

13-7 Assume a frictionless surface for block A in Figure P13-7 and neglect the inertia of the pulley. Determine the acceleration of block A.

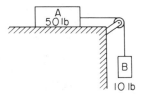

Figure P13-7

13-8 Neglect the inertia of the pulley and determine the acceleration of masses A and B in Figure P13-8.

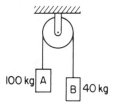

Figure P13-8

13-9 Neglect pulley inertia and determine the acceleration of (a) mass A, and (b) mass B in Figure P13-9.

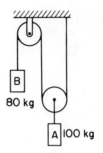

Figure P13-9

13-10 Neglect pulley inertia, and with the system initially at rest, determine the distance that mass B in Figure P13-10 will move in 4 s.

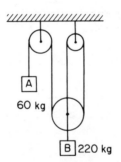

Figure P13-10

13-11 The system shown in Figure P13-11 is released from rest. Determine the acceleration of each mass. (Neglect beam inertia.)

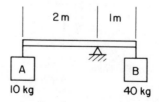

Figure P13-11

13-12 Neglect pulley inertia and determine the acceleration of mass B in Figure P13-12.

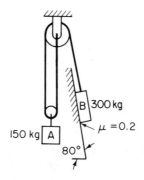

Figure P13-12

13-13 Block A has a mass of 53.1 kg; block B has a mass of 13.3 kg. Determine the velocity of B at $t=2$ s if the system is initially at rest and is released from the position shown in Figure P13-13.

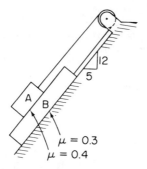

Figure P13-13

13-14 The truck in Figure P13-14 weighs 3500 lb, and the crate on it weighs 1500 lb. Assume equal weight distribution on all wheels and determine (a) the maximum deceleration rate of the loaded truck (assume that the crate does not slide) and (b) the maximum deceleration before the crate does slide.

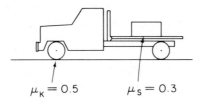

Figure P13-14

13-15 Starting from rest, weight A in Figure P13-15 is subject to $P=100$ lb for 5 sec. Determine the velocity of weight A at $t=5$ sec.

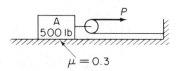

Figure P13-15

13-16 A 4000-lb car travelling at 60 mph decelerates at a constant rate and comes to a stop over a distance of 200 ft. Determine the minimum coefficient of friction between the pavement and the car's tires.

13-17 A car comes to rest in a controlled skid from a speed of 120 km/h. The skid length measures 220 m. The car mass is 900 kg. Determine the average friction force applied by the road on the tires during the skid.

13-18 A car skids to rest from a velocity of 50 mph. If the coefficient of static friction is 0.56, determine the length of skid.

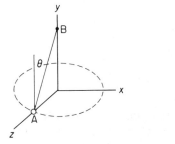

Figure P13-19

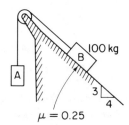

Figure P13-20

13-19 A 4 kg ball fastened to a cord swings in a horizontal circle with a 2 m diameter with a velocity of 2.5 m/s (Figure P13-19). Determine (a) the tension in the cord; (b) the angle θ; and (c) the length of cord AB.

13-20 The coefficient of kinetic friction for mass B in Figure P13-20 is 0.25. Determine the acceleration of mass A if it has a mass of (a) 30 kg, and (b) 50 kg.

13-21 A horizontal disk accelerates from rest at 5 rad/s². The coefficient of static friction between it and a 2.5 kg block is 0.20. At what angular speed of the disk will the block begin to slide (Figure P13-21)?

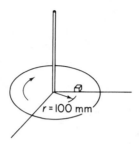

Figure P13-21

13-22 An amusement ride consists of a large horizontal circular platform turning at 9 rpm. People attempt to sit on this platform without slipping toward the outer edge. The coefficient of static friction is 0.15. Determine the distance from the center of the platform that a 120-lb person may sit before slipping occurs. Does the weight of a person have any effect on this distance?

13-23 A 40-kg mass (A) rests on a horizontal platform that revolves about a vertical axis (Figure P13-23). Masses A and B are connected by a cord as shown. If mass A begins to slip when the platform has a speed of 2 rad/s, determine the mass of B. Neglect rolling friction of B.

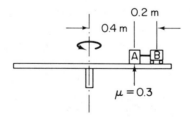

Figure P13-23

13-24 Two trailers hooked together, each weighing 1000 lb, are accelerated to the right by a force of 500 lb. Each trailer has a constant rolling resistance force of 100 lb. If the trailers start from rest, how far do they travel in 15 seconds?

13-25 Blocks A and B, when together as shown in Figure P13-25, have an initial velocity of 2.5 m/s to the right and slide to rest due to friction. How far do they slide?

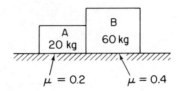

Figure P13-25

13-26 Crate A in Figure P13-26 is given an initial velocity of 6 ft/sec down the inclined plane. If the coefficient of kinetic friction is 0.5, how far will the crate slide down the plane before coming to a stop?

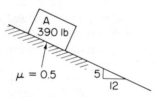

Figure P13-26

13-27 Block A in Figure P13-27 is given an initial velocity of 6 m/s up the incline. It comes to a stop in distance d and then slides back down the incline with uniform acceleration. Determine (a) distance d, and (b) the velocity of A when it returns to its original point.

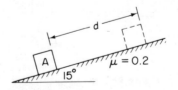

Figure P13-27

13-28 It takes 10 s to accelerate a wheel from rest to 900 rpm. If the wheel has a mass moment of inertia of 2 kg·m², determine the constant torque required to produce this acceleration.

13-29 A flywheel is accelerated from rest to 500 rpm by an applied torque of 150 lb-ft. If it has a mass moment of inertia of 100 ft-lb-sec², determine the time required.

13-30 A torque of 160 lb-ft is applied to a large propeller fan for 7 sec causing it to accelerate from rest to 300 rpm. Determine the mass moment of inertia about its center of rotation.

13-31 Cylinder A has a mass of 25 kg and a radius of gyration of $k=0.5$ m. At the instant shown in Figure P13-31, it is turning clockwise and is being braked by the application of a force, P, of 40 N. Determine the angular deceleration of the wheel.

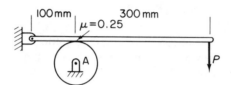

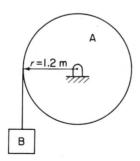

Figure P13-31

13-32 Cylinder A in Figure P13-32 has a mass moment of inertia about its axis of rotation of 40 kg·m². Mass B of 8 kg is suspended by a cable wrapped around A. Determine the angular acceleration of A and the linear acceleration of B if mass B is allowed to drop from rest.

Figure P13-32

13-33 A 50-lb sphere with a radius of gyration of 0.4 ft is accelerated about its centroidal axis. Determine the torque required to accelerate it from rest to 800 rpm in 15 seconds.

13-34   A 100-lb wheel, 2 ft in diameter, is accelerated from rest by weight A moving downward (Figure P13-34). Wheel motion is retarded by a bearing friction moment of 20 lb-ft and a braking force due to a spring force of 80 lb. How far will A fall in 6 seconds? (Assume $I = 1/2\,mr^2$ for B.)

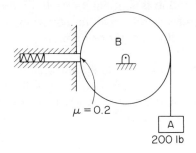

Figure P13-34

13-35–13-38   The two pulleys fastened together in Figures P13-35 to P13-38 have diameters of 1.5 and 1 m. Their combined mass moment of inertia is 35 kg·m². Determine (a) the angular acceleration of the pulleys, and (b) the tension T in the cord supporting the 612 kg mass when it is released from rest.

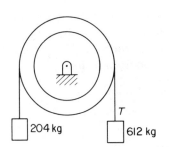

Figure P13-35

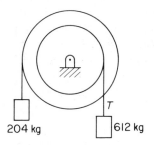

Figure P13-36

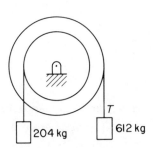

Figure P13-37

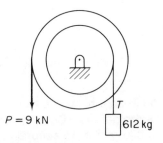

Figure P13-38

13-39 The mechanism shown in Figure P13-39 is started under fully loaded conditions, which requires a torque of 10 lb-ft. When the mechanism is started, it has a mass moment of inertia of 1.66 ft-lb-sec². How long will it take the drive pulley to accelerate from rest to its full speed of 90 rpm if the belt tensions applied to it are 70 lb and 30 lb?

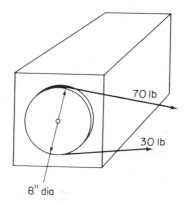

Figure P13-39

13-40 For gear A, $I=0.4$ kg·m²; for gear B, $I=1.2$ kg·m² (Figure P13-40). If B is to be accelerated at 10 rad/s² clockwise, what torque must be applied to gear A?

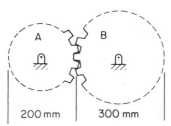

Figure P13-40

13-41 The long slender rod in Figure P13-41 has a mass of 4.5 kg and, while at rest, is acted upon by a force, $F=90$ N. Determine the reaction components at A.

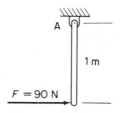

Figure P13-41

13-42 The solid cylinder in Figure P13-42 weighs 96.6 lb and, while at rest, is acted upon by a force, $F=60$ lb. Determine (a) the reaction components at A, and (b) the angular acceleration of the cylinder about A.

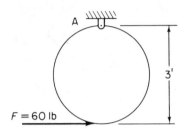

$F=60$ lb

Figure P13-42

13-43 The hollow cylinder in Figure P13-43 has a mass of 14 kg and, while at rest, is acted upon by a force, $F=350$ N. Determine (a) the reaction components at A, and (b) the angular acceleration of the cylinder about A.

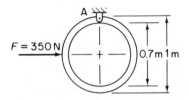

$F=350$ N

0.7 m 1 m

Figure P13-43

13-44 What force P is required to accelerate the 4000-lb sewer pipe in Figure P13-44 at 1 ft/sec² to the right?

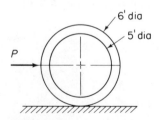

6' dia

5' dia

P

Figure P13-44

13-45 A lawn roller 0.7 m in diameter has a mass of 22 kg. Neglect rolling resistance and determine the force P required to accelerate the roller from rest to 5 m/s in a distance of 3 m (Figure P13-45).

$$(I_c = 10 \text{ kg} \cdot \text{m}^2.)$$

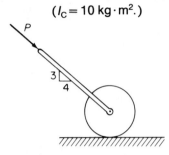

Figure P13-45

13-46 The rotation of winch drum A in Figure P13-46 causes motion of weight B to the right. The winch drum has a mass moment of inertia of 10 ft-lb-sec². Determine the torque that must be applied to the drum to cause weight B to accelerate at 4 ft/sec².

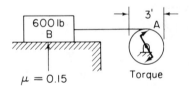

Figure P13-46

13-47 The cylinder and hub in Figure P13-47 have a total mass of 30 kg and a radius of gyration of 0.5 m. Assume no slippage. Determine the acceleration of the mass center of the cylinder and hub when $P=90$ N.

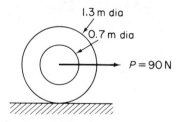

Figure P13-47

13-48 The same as Problem 13-47—except *P* is applied as shown in Figure P13-48.

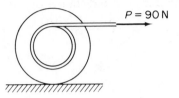

$P = 90\,\text{N}$

Figure P13-48

13-49 The same as Problem 13-47—except *P* is applied as shown in Figure P13-49.

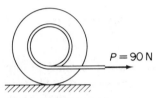

$P = 90\,\text{N}$

Figure P13-49

13-50 The same as Problem 13-47—except *P* is applied as shown in Figure P13-50.

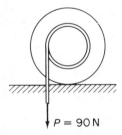

$P = 90\,\text{N}$

Figure P13-50

13-51 Cylinder A in Figure P13-51 weighs 322 lb. Neglect the weight of the hub. The coefficient of static friction is 0.2. Determine the maximum force *P* allowable before slippage occurs at the hub. What is the maximum acceleration of the mass center of A?

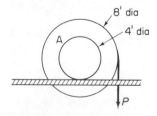

8' dia

4' dia

A

P

Figure P13-51

13-52 Neglect pulley friction and inertia and determine the acceleration of the mass center of cylinder A in (Figure P13-52) when the system is released from rest.

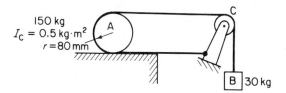

Figure P13-52

13-53 Cylinder A in Figure P13-53 has a diameter of 0.6 m and a mass of 260 kg. Assume that there is no slippage of A and that mass B is released from rest; determine (a) the tension in the rope, and (b) the distance that B will drop in 20 s. (Neglect the mass and inertia of the rope and pulley.)

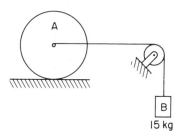

Figure P13-53

13-54 A cable is unwound from a reel by anchoring the cable and moving to the left with the reel and its carriage. At the instant shown in Figure P13-54, the reel weighs 3220 lb and has a radius of gyration of 1.8 ft. The reel carriage weighs 1000 lb. Neglect the angular inertia of the carriage wheels and the friction at the axle of the reel. Assume a constant rolling resistance of the carriage of 100 lb. Determine the acceleration of the carriage when a force, $P = 600$ lb, is applied.

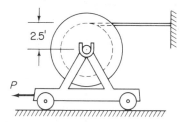

Figure P13-54

13-55 A 64.4-lb sphere with an 8-in. diameter is released from rest on a slope that is inclined at 60° to the horizontal. Determine the minimum coefficient of friction that it must have in order for it to roll without slipping when released.

13-56 A cylinder 1.2 m in diameter, having a mass of 40 kg and a rope wrapped around it (Figure P13-56) is released from rest. Determine the velocity of the mass center of the cylinder after it has dropped 2.5 m.

Figure P13-56

13-57 The 18-kg cylinder in Figure P13-57, when released from rest, slips on the inclined surface. Determine the angular acceleration of the cylinder $(I_c = 1.84 \text{ kg} \cdot \text{m}^2)$.

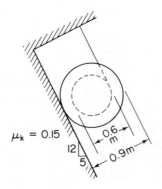

$\mu_k = 0.15$

12
5

0.6 m
0.9 m

Figure P13-57

14

WORK, ENERGY, AND POWER

14-1   INTRODUCTION

In Chapter 13, when we made use of the inertia-force method, we were concerned with force and acceleration. If distance and velocity were further required, we would employ one or two of the three basic kinematic equations (Section 10-5).

The work-energy method gives us a direct measure of distance and velocity values. If acceleration is required, the three main kinematic equations are used again.

In the previous chapter, we looked at bodies or systems of bodies in motion. They were in motion due to an unbalance of forces acting on them. A body in motion possesses *energy*, receiving it from a force *F* acting on it through a distance *s*. This quantity of energy we call *work*. We now come to a method that is an accounting process for all quantities of energy. According to the law of the conservation of energy, energy cannot be lost—merely converted from one form to another. Three common types of energy are work, potential energy, and kinetic energy. One or two of these types of energy may initiate motion from one point to another and be converted to another form of energy in the process. Our equations will account for all these energies.

14-2   WORK OF A CONSTANT FORCE

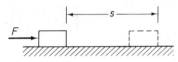

Figure 14-1

Work (denoted as U) is due to an applied force's acting over some distance. The most familiar work would be that shown in Figure 14-1, wherein a force F moves an object a distance s.

$$\text{work} = \text{force} \times \text{distance}$$
$$U = F \times s$$

where:

U = work in joules (1 J = 1 N·m)
F = force in newtons (N)
s = distance in meters (m)

or, in the English System:

U = work in ft-lb
F = force in lb
s = distance in ft

In the SI metric system, with work expressed in units of joules and torque in units of newton meters (N·m), the units provide more of a differentiation between the terms than is provided by the terms in the English system (work in ft-lb and torque in lb-ft).

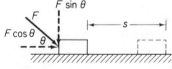

Figure 14-2

If the force is applied as in Figure 14-2, then $U = F\cos\theta \times s$ since only the horizontal component causes motion in the direction in which distance s is measured. There is no vertical movement, so $F\sin\theta$ does no work.

In these two examples, you will have noted that the force and distance are in the same direction and that the force is constant.

Example 14-1

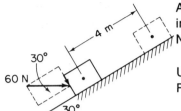

Figure 14-3

An 8-kg block is pushed 4m up a 30° inclined plane by a horizontal force of 60 N. Determine the work done by this force.

Using the forces and the distance shown in Figure 14-3, we get:

$$U = (60 \cos 30) \times 4$$
$$= 52 \times 4$$
$$U = 208 \text{ J}$$

14-3 WORK OF A VARIABLE FORCE

Work is not always due to a constant force acting through a distance. There are many practical applications of work resulting from a force that varies as it moves through distance s. To describe some of the more complex variations of force requires either a calculus approach or the drawing of nonlinear curves on a force-displacement diagram. We will not discuss these more complex variable forces but will limit ourselves to forces that vary in a linear fashion.

The force required to stretch or compress most springs is a force that varies linearly. For example, if a force of 10 lb compresses a spring 2 in. long, then a force of 30 lb compresses that spring 6 in. This relationship expressed in equation form is:

$$F = kx \tag{14-1}$$

F = the force exerted on the spring in lb
k = the spring constant in lb/in.
x = the change in length of the spring, in in., as measured from its unloaded or free length

The corresponding SI units are:

F = spring force in newtons
k = spring constant in netwons/meter (N/m)
x = change in length in meters (m)

Example 14-2

What is the spring constant of a spring that is compressed 50 mm by a force of 25 N? What force is required to compress it a total of 125 mm?

$$F = kx$$
$$25 = k \times 0.05 \quad \text{(using length in meters)}$$
$$\underline{k = 500 \text{ N/m}}$$

$$F = kx$$
$$= 500 \times 0.125$$
$$\underline{F = 62.5 \text{ N}}$$

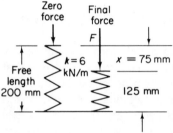

Figure 14-4

Consider now the work required to compress the spring in Figure 14-4. The spring is compressed a distance x. The force acting through this distance is varied from 0 to the final force F. Therefore, the average force over this distance is $F/2$, but $F = kx$ and $F/2 = kx/2$. Substituting into the equation

$$\text{work} = \text{force} \times \text{distance}$$

we have:

$$\text{spring work} = \left(\frac{kx}{2}\right) \times (x)$$

$$U = \frac{1}{2} kx^2 \qquad (14\text{-}2)$$

Using the values in Figure 14-4, we get:

$$U = \frac{1}{2} \times 6000(0.075)^2$$

$$U = 16.9 \text{J}$$

Another way of expressing this equation and its corresponding calculations is on a work diagram such as Figure 14-5, where the area on the diagram is equal to the work.

For larger springs, the spring constant may be kN/m. While units of meters may seem awkward to use for a spring that has a

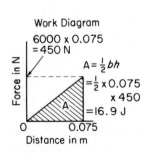

Figure 14-5

deflection of only a few mm, meters are consistent with the SI convention and will yield spring work (Equation 14-2) in joules.

Example 14-3

A spring has a spring constant of 80 lb/in. and a free length of 10 in. The spring is stretched to point A, where its total length is 13 in. What work would now be required to stretch it to point B, where it would be 15 in. in length?

The spring is stretched 3 in. and then 5 in. beyond its free length. The work required to stretch the spring from 3 to 5 in. is the difference between the work to stretch it 5 in. and the work required to stretch it 3 in. or:

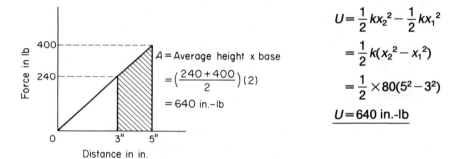

$$U = \frac{1}{2}kx_2{}^2 - \frac{1}{2}kx_1{}^2$$

$$= \frac{1}{2}k(x_2{}^2 - x_1{}^2)$$

$$= \frac{1}{2} \times 80(5^2 - 3^2)$$

$$U = 640 \text{ in.-lb}$$

A = Average height x base

$$= \left(\frac{240 + 400}{2}\right)(2)$$

$$= 640 \text{ in.-lb}$$

Figure 14-6

This work value could also have been shown on the work diagram, Figure 14-6.

Note that the units of work are in.-lb and may often have to be converted to ft-lb in further problem calculations.

14-4 POTENTIAL AND KINETIC ENERGY-TRANSLATIONAL

An object lowered from one level to another has the ability or potential to do work since it has a force (weight) acting through a distance (height). For this reason, the object is said to have *potential energy* (PE): it has the potential either to do work or to convert its energy into another form, such as heat or motion.

$$PE = W \times h$$

where:

PE = potential energy in joules (J)
W = force of gravity (or weight) in N
h = vertical height in m

and, in the English system:

PE = potential energy in ft-lb
W = weight in lb
h = vertical height in ft

The reference level at which we measure height is arbitrary since we are only concerned with a change in height or potential energy. For this reason, if an object moves from one level to another, we often use either one of these levels as a reference datum.

When an object moves up against the weight, the weight's potential energy increases. As the weight falls, it loses potential energy.

Example 14-4

A 1000-lb elevator moves upward from the 10th floor to the 14th floor, a distance of 44 ft. What is the increase in the potential energy of the elevator?

Using the 10th floor as the reference datum from which we measure height, we have:

$$PE = W \times h$$
$$= 1000 \times 44$$
$$\underline{PE = 44,000 \text{ft-lb}}$$

The potential energy of the elevator increased by 44,000 ft-lb; therefore, a corresponding amount of work must have been done to raise the elevator to the reference level.

Example 14-5

While driving through a valley, a 1600-kg car has a resulting elevation drop of 400 m. Determine its decrease in potential energy.

The force of gravity or weight:

$$W = 1600 \text{ kg} \times 9.81 \text{ m/s}^2$$
$$W = 15.7 \text{ kN}$$
$$PE = W \times h$$
$$= (15.7 \times 10^3) \times 400$$
$$= 6{,}280{,}000 \text{ J}$$
$$PE = 6.28 \text{ MJ}$$

Work or energy is required to start an object moving—and to stop a moving object as well. The energy of a moving object is *kinetic energy*.

Consider now the work required to accelerate an object from an initial velocity v_0 to some final velocity v.

$$U = F \times s \quad \text{and} \quad F = ma$$

therefore:

$$U = mas \tag{1}$$

but

$$v^2 = v_0^2 + 2as \quad \text{or} \quad a = \frac{v^2 - v_0^2}{2s} \tag{2}$$

Substituting Equation 2 into Equation 1, we get:

$$U = m\left(\frac{v^2 - v_0^2}{2s}\right) s$$

$$U = \frac{1}{2} m (v^2 - v_0^2)$$

The kinetic energy change for a given speed change is:

$$KE = \frac{1}{2} m(v^2 - v_0^2) \tag{14-3}$$

For an initial velocity of zero:

$$KE = \frac{1}{2} mv^2$$

where, in the SI system:

$$KE = \text{kinetic energy in J}$$
$$m = \text{mass in kg}$$
$$v = \text{velocity in m/s}$$

and, in the English system:

$$KE = \text{kinetic energy in ft-lb}$$
$$m = \text{mass in slugs}$$
$$v = \text{velocity in ft/sec}$$

This equation does not apply to rotational motion; it applies only to translational motion. (Translational motion may be either rectilinear or curvilinear.)

Example 14-6

A "slap shot" in the game of hockey can cause an increase in the velocity of a 5-oz puck from 10 mph to 100 mph (60 mph = 88 ft/sec). What would be the corresponding increase in the kinetic energy of the puck?

$$v_0 = \frac{10}{60} \times 88 = 14.67 \text{ ft/sec}$$

$$v = 146.7 \text{ ft/sec}$$

$$KE = \frac{1}{2} m(v^2 - v_0^2)$$

$$= \frac{1}{2} \times \frac{5/16}{32.2} \left[(146.7)^2 - (14.67)^2 \right]$$

$$KE = 100 \text{ ft-lb}$$

Example 14-7

A 100-g sample in a centrifuge is rotated at a speed of 3600 rpm. If the sample is 200 mm from the center, what is its kinetic energy?

$$v = r\omega$$

$$= 0.2 \times \frac{2\pi \times 3600}{60}$$

$$v = 75.4 \text{ m/s}$$

$$KE = \frac{1}{2} mv^2$$

$$= \frac{1}{2} \times 0.1 \times (75.4)^2$$

$$KE = 284 \text{ J}$$

14-5   CONSERVATION OF ENERGY—TRANSLATIONAL

The mechanical forms of energy that we have looked at are work, potential energy, and kinetic energy. Another form of energy, perhaps not so obvious, is the work of the force of friction acting through a distance. This energy is, in turn, dissipated as heat.

By *conservation of energy*, we mean that, no matter what motion a given system has, the total initial energy equals the total final energy; no matter what form the energy takes or changes to, all quantities of energy must be accounted for. With the accounting system that we will use here, we can handle systems that are decelerating, accelerating, and at constant velocity; we will even be able to deal with the change of position of a system of objects that has both initial and final velocities of zero.

The method consists of considering only initial and final conditions and writing an energy equation—the intermediate values or conditions are of no concern. The first step of our method of attack will be to analyze the motion of the system and to ask ourselves, "What is the change in energy that causes the resulting motion?" There is some object or portion of a system that causes motion; the energy of this portion is converted into other forms of energy. We, therefore, equate this activating energy with all the other amounts of energy into which it has been converted in the system.

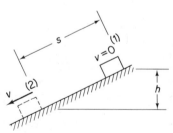

Figure 14-7

For example, consider a block that slides down a plane and reaches the bottom with some velocity, *v* (Figure 14-7). The amount of energy that caused this motion is accounted for by the loss of potential energy of the block since the weight acted vertically over a height, *h*. This energy was converted to or reappears as kinetic energy at point (2); in addition, the work of the friction force dissipated as heat over distance, *s*.

$$\Delta PE = \Delta KE + \text{friction work}$$

Example 14-8

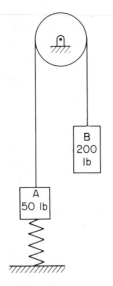

Figure 14-8

For the system shown in Figure 14-8, B weighs 200 lb and A weighs 50 lb. The spring is at its free length and has a spring constant of 25 lb/ft. If the system is initially at rest and the weight and inertia of the cable and pulley are neglected, determine the velocity of weight B after it has moved 3 ft.

The activating energy that causes the motion results from the loss of potential energy of B. Equating this energy with all the other forms of energy into which it is dispersed, we have:

activating energy = resulting energies

PE loss of B = KE gain of B + KE gain of A + PE gain of A + spring work.

$$W_B \times h = \left(\frac{1}{2} m_B v^2\right) + \left(\frac{1}{2} m_A v^2\right)$$

$$+ (W_A \times h) + \left(\frac{1}{2} ks^2\right)$$

$$200 \times 3 = \left(\frac{1}{2} \times \frac{200}{32.2} v^2\right) + \left(\frac{1}{2} \times \frac{50}{32.2} v^2\right)$$

$$+ (50 \times 3) + \left[\frac{1}{2} \times 25 \times (3)^2\right]$$

$$\underline{v = 9.33 \text{ ft/sec} \downarrow}$$

You may have noted that the answer was expressed in ft and sec since the units of *g* are ft/sec².

Example 14-9

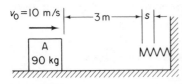

Figure 14-9

Block (A) in Figure 14-9 has a mass of 90 kg and an initial velocity of 10 m/s to the right. The spring constant is 12kN/m, and the coefficient of kinetic friction is 0.15. How much will the spring deflect in bringing the block to rest?

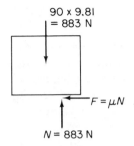

90×9.81
$= 883\ N$

$F = \mu N$

$N = 883\ N$

Figure 14-10

From Figure 14-10, the force of friction is:

$$F = \mu N$$
$$F = 0.15 \times 883 = 132\ N$$

Activating energy is equal to resulting energy.

KE loss of A = friction work + spring energy

$$\frac{1}{2}mv^2 = F(3+s) + \frac{1}{2}ks^2$$

$$\frac{1}{2} \times 90 \times (10)^2 = 132(3+s) + \frac{1}{2} \times 12{,}000\,s^2$$

$$45.45s^2 + s - 31.1 = 0$$

$$s = \frac{-1 \pm \sqrt{(1)^2 - (4)(45.45)(-31.1)}}{2 \times 45.45}$$

$$s = 0.816\ m$$

Example 14-10

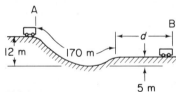

A

$12\ m$ $170\ m$

$5\ m$

Figure 14-11

A 1300-kg car starts from rest at A, coasts 170 m through a dip, and comes to a stop at B after coasting a distance d on level ground (Figure 14-11). If the rolling resistance is a constant force of 220 N, determine the distance d.

The kinetic energy is zero at both initial and final conditions, and we do not have to be concerned with the velocity reached at some intermediate stage. What we do have to realize is that a loss of potential energy was the activating energy; the potential energy was dispersed as work, overcoming the rolling resistance force.

PE loss = work of rolling resistance

$$1300 \times 9.81(12-5) = (170+d)220$$

$$\underline{d = 236\ m}$$

14-6   KINETIC ENERGY—ANGULAR

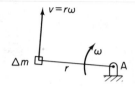

Figure 14-12

Our equation for rectilinear kinetic energy is $KE = 1/2 mv^2$. We will now convert this equation into angular terms by considering Figure 14-12. There, a mass m rotates about point A with an angular velocity ω. Considering mass Δm at a distance r from the center of rotation A, we can write the kinetic energy equation as:

$$KE = \frac{1}{2}\Delta m (r\omega)^2$$

The kinetic energy of the total mass m is:

$$KE = \Sigma \frac{1}{2} mr^2 \omega^2$$

$$KE = \frac{\omega^2}{2} \Sigma mr^2$$

But, as we saw in Section 9-5, mass moment of inertia is given by:

$$I = \Sigma mr^2$$

Therefore, for angular kinetic energy:

$$KE = \frac{1}{2} I \omega^2 \qquad\qquad (14\text{-}4)$$

where, in the SI system:

  KE = kinetic energy in J

  ω = angular velocity in rad/s

  I = mass moment of inertia about the center of rotation in kg·m²

and, in the English system:

  KE = kinetic energy in ft-lb

  ω = angular velocity in rad/sec

  I = mass moment of inertia about the center of rotation in slugs-ft² or ft-lb-sec²

Tables and formulae for mass moment of inertia are usually set up, taking as their reference the center of mass of the object. For cases in which the center of mass and the center of rotation are not coincident, we will consider the total kinetic energy as being equal to the angular KE, using the mass moment of inertia (I_c) about the center of mass, plus the rectilinear KE of the center of mass. Section 14-8 will cover this in more detail; examples will show what we mean by this.

Example 14-11

A 96.6-lb rotor is rotated at 1500 rpm while on a dynamic balancing machine. Assume the rotor to be a cylinder with a radius of 6 in. and determine its kinetic energy at this speed.

From Table 9-2:

$$I_c = \frac{1}{2}mr^2$$

$$= \frac{1}{2} \times \frac{96.6}{32.2} 0.5^2$$

$$I_c = 0.375 \text{ ft-lb-sec}^2$$

$$KE = \frac{1}{2}I_c\omega^2$$

$$= \frac{1}{2} \times 0.375 \times \left(\frac{1500 \times 2\pi}{60}\right)^2$$

$$\underline{KE = 4620 \text{ ft-lb}}$$

Example 14-12

A 150-mm diameter shaft is being turned on a lathe at 80 rpm. If the shaft mass is 210 kg, determine its kinetic energy.

From Table 9-2:

$$I_c = \frac{1}{2}mr^2$$

$$= \frac{1}{2} \times 210 \times (0.075)^2$$

$$I_c = 0.591 \text{ kg} \cdot \text{m}^2$$

$$KE = \frac{1}{2} I_c \omega^2$$

$$= \frac{1}{2} \times 0.591 \times \left(\frac{80 \times 2\pi}{60} \right)^2$$

$$KE = 20.7 \text{ J}$$

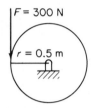

Figure 14-13

Another form of energy often encountered in rotational motion is that due to a torque's being applied during a given number of revolutions. In Figure 14-13, a force of 300 N is applied to a wheel with a radius of 0.5 m as the wheel turns through three revolutions. The work could be calculated in one of two ways:

1. $U = F \times s$

   $= F \times (3 \times \text{circumference})$

   $= 300(3 \times 2\pi \times 0.5)$

   $\underline{U = 2830 \text{ J}}$

2. $U = F \times s, \quad \text{but } s = r\theta$

   $U = Fr\theta \text{ but torque } T = F \times r$

   $U = \text{torque} \times \theta$

where:

torque is in N·m (or lb-ft)
s = distance in radians
U = work in J (or ft-lb)

In this case:

$$U = (300 \times 0.5)(3 \times 2\pi)$$
$$\underline{U = 2830 \text{ J}}$$

Example 14-13

A torque of 250 in.-lb is transmitted by a shaft rotating at 800 rpm. What is the work in ft-lb transmitted by the shaft in one minute?

$$U = \text{torque} \times \theta$$

$$= \frac{250}{12} \times 800 \times 2\pi$$

$$\underline{U = 104{,}700 \text{ ft-lb}}$$

14-7 CONSERVATION OF ENERGY—ANGULAR

We must account for all energies in rotational motion—just as we did in translational motion.

Example 14-14

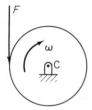

Figure 14-14

A wheel with a radius of 15 in. ($I_c = 0.8$ ft-lb-sec^2) is braked by a force applied at its outer diameter (Figure 14-14). If the wheel speed is reduced from 6 rad/sec to 2 rad/sec as it turns three revolutions, determine the braking force.

The activating energy is accounted for by the loss of kinetic energy of the wheel, and this energy is completely absorbed by the braking force.

$$\frac{1}{2} I(\omega_2{}^2 - \omega_1{}^2) = \text{torque} \times \theta$$

$$\frac{1}{2} \times 0.8(6^2 - 2^2) = \left(F \times \frac{15}{12} \right)(3 \times 2\pi)$$

$$\underline{F = 0.544 \text{ lb}}$$

Example 14-15

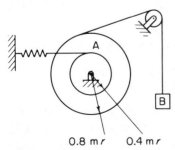

0.8 m r   0.4 m r

Figure 14-15

Body A in Figure 14-15 has a moment of inertia about its center of mass of 55 kg·m^2. Block B has a mass of 50 kg and a velocity of 2.7 m/s. The spring has a modulus of 400 N/m and is stretched 0.2 m at the instant shown. Determine (a) the velocity of B after it has dropped 0.7 m, and (b) the maximum velocity reached by block B.

(a) The activating energy or the energy input to the system is the potential energy loss of B.

PE loss of $B = \Delta KE_B + \Delta KE_A + $ spring work

$$W \times h = \frac{1}{2} m(v_2^2 - v_1^2) + \frac{1}{2} I(\omega_2^2 - \omega_1^2) + \frac{1}{2} k(s_2^2 - s_1^2)$$

where:

$$v = r\omega = 0.8\omega \quad \text{or} \quad \omega = \frac{v}{0.8} \text{ for A}$$

and

final spring distance = initial distance + proportion of block B distance

$$= 0.2 + \frac{0.4}{0.8} \times 0.7$$

final spring distance = 0.55 m

Filling known information into our energy equations, we get:

$$(50 \times 9.81 \times 0.7) = \left[\frac{1}{2} \times 50(v_2^2 - 2.7^2) \right] + \frac{1}{2} \times 55 \left[\left(\frac{v_2}{0.8} \right)^2 - \left(\frac{2.7}{0.8} \right)^2 \right]$$

$$+ \frac{1}{2} \times 400 \left[(0.55)^2 - (0.2)^2 \right]$$

$$v_2 = 3.4 \text{ m/s} \downarrow$$

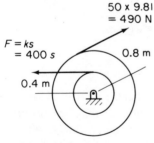

50 x 9.81
= 490 N

$F = ks$
= 400 s

0.8 m

0.4 m

Figure 14-16

(b) As long as block B causes an unbalanced clockwise torque on body A, its velocity will increase. Block B will reach maximum velocity when the spring force is sufficient to provide a balancing counterclockwise torque (Figure 14-16).

$$490 \times 0.8 = 0.4 \times 400 s$$

$$s = 2.45 \text{ m}$$

The spring stretches a total of 2.45 m, and block B drops 4.9 m. From the initial position shown, spring distance = 2.45 − 0.2 = 2.25 m and block B travels a distance of 4.9 − 0.7 = 4.2 m. Block B would continue to drop beyond this level due to inertia, but it would be decelerating.

Writing the energy equation as before, we have:

PE loss of $B = \Delta KE_B + \Delta KE_A +$ spring work

$$490 \times 4.2 = \frac{1}{2} \times 50 [(v)^2 - (2.7)^2] + \frac{1}{2} \times 55 \left[\left(\frac{v}{0.8}\right)^2 - \left(\frac{2.7}{0.8}\right)^2 \right]$$

$$+ \frac{1}{2} \times 400 [(2.45)^2 - (0.2)^2]$$

$$\underline{v = 4.47 \text{ m/s}\downarrow}$$

The maximum velocity of B is thus 4.47 m/s.

14-8 CONSERVATION OF ENERGY—PLANE MOTION

Plane motion consists of both translational motion and rotational motion. The problems that we will solve are very similar to those in Section 13-5—but we will now use the energy method.

Example 14-16

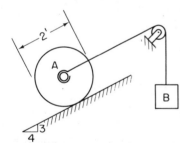

Cylinder A in Figure 14-17 weighs 200 lb and has a mass moment of inertia about its center of mass of 3 ft-lb-sec². Block B weighs 100 lb. The weight and inertia of the pulley and cable can be neglected. If the system starts from rest, determine the velocity of B after A has rolled 10 ft on the slope.

Figure 14-17

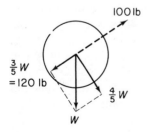

The first point to check is whether cylinder A rolls up or down the slope. As shown by Figure 14-18, the component (3/5 W) of A is greater than the cable tension, so A rolls down the slope.

Figure 14-18

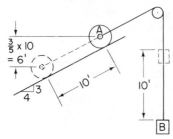

$\frac{3}{5} \times 10$
$= 6'$

Figure 14-19

Now obtain the distances travelled (Figure 14-19). Equate the activating energy with all of the final forms of energy. The potential energy loss of A results in:

1. KE of B
2. PE increase of B
3. KE of A (rectilinear)
4. KE of A (angular)

$$\Delta PE_A = \Delta KE_B + \Delta PE_B + \Delta KE_{A(rect.)} + \Delta KE_{A(ang.)}$$

$$W_A h_A = \left(\frac{1}{2} mv^2\right) + (W_B \times h_B) + \left(\frac{1}{2} mv^2\right)\left(\frac{1}{2} I_c \omega^2\right)$$

where v is the rectilinear velocity of A and B.

$$\omega = \frac{v}{r} = \frac{v}{1}$$

$$200 \times 6 = \left(\frac{1}{2} \times \frac{100}{32.2} v^2\right) + (100 \times 10) + \left(\frac{1}{2} \times \frac{200}{32.2} v^2\right) + \left(\frac{1}{2} \times 3 \times v^2\right)$$

$$\underline{v = 5.7 \text{ ft/sec}\uparrow}$$

Example 14-17

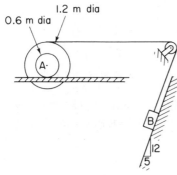

Figure 14-20

Cylinder A rolls in a slot and on a hub 0.6 m in diameter (Figure 14-20). A cable is wrapped around the cylinder's 1.2 m diameter. Cylinder A has a mass of 70 kg and a radius of gyration of 0.5 m with respect to the center of mass. Mass B is 26.5 kg and has a coefficient of kinetic friction of 0.2. Assume that there is no slippage of A and that the system is initially at rest. Determine the angular velocity of A after B has slid 2 m along the inclined plane.

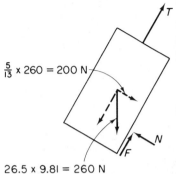

$\frac{5}{13}$ x 260 = 200 N

26.5 x 9.81 = 260 N

Figure 14-21

There is no question as to the direction of motion, but a free-body diagram of B is required in order to obtain the friction force (Figure 14-21).

$$N = 200 \text{ N}$$

and

$$F = \mu N = 0.2 \times 200 = 40 \text{ N}$$

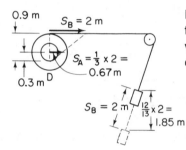

0.9 m

$S_B = 2$ m

$S_A = \frac{1}{3} \times 2 = 0.67$ m

0.3 m D

$S_B = 2$ m $\frac{12}{13} \times 2 = 1.85$ m

Figure 14-22

Next, referring to Figure 14-22, we determine some of the various distances involved. Note that D is the instantaneous center of rotation of A.

Since we are solving for the angular velocity of A, all velocities could be expressed in terms of ω (Figure 14-23).

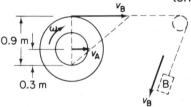

v_B

0.9 m

ω

v_A

0.3 m

v_B

Figure 14-23

$$v_A = r\omega = 0.3\omega$$
$$v_B = 3v_A = 0.9\omega$$

The radius of gyration is:

$$k = \sqrt{\frac{I}{m}}$$

or

$$I = k^2 m$$
$$= (0.5)^2 \times 70$$
$$I = 17.5 \text{ kg} \cdot \text{m}^2$$

The potential energy loss of B results in:

1. KE of B
2. friction work on B
3. KE of A (angular)
4. KE of A (rectilinear)

$$PE_B = KE_B + \text{friction work} + KE_{A(\text{ang.})} + KE_{A(\text{rect.})}$$

$$W \times h = \left(\frac{1}{2} mv_B{}^2\right) + (F \times s) + \left(\frac{1}{2} I_\omega{}^2\right) + \left(\frac{1}{2} mv_A{}^2\right)$$

$$260 \times 1.85 = \left[\frac{1}{2} \times 26.5(0.9\omega)^2\right] + (40 \times 2)$$

$$+ \left(\frac{1}{2} \times 17.5\omega^2\right) + \left[\frac{1}{2} \times 70(0.3\omega)^2\right]$$

$$\underline{\omega = 4.21 \text{ rad/s } \downarrow}$$

14-9   POWER AND EFFICIENCY

We have examined the various forms of energy—work being one of them. We will now concern ourselves not simply with the quantity of work but also with the time required to accomplish this work. The rate of doing work is *power*.

$$\text{power} = \frac{\text{work}}{\text{time}}$$

$$P = \frac{U}{t} \qquad (14\text{-}6)$$

where, in the SI system:

U = work in joules (J)
t = time in seconds (s)
P = power in watts (W), since 1 watt is
   defined as the rate of work of 1 joule per
   second, 1 W = 1 J/s = $\dfrac{\text{N} \cdot \text{m}}{\text{s}}$

and, in common units for the English system:

U = work in ft-lb
t = time in sec
P = power in ft-lb/sec

since

$$1 \text{ horsepower} = 550 \text{ ft-lb/sec}$$
$$1 \text{ horsepower} = 33{,}000 \text{ ft-lb/min}$$
$$\text{Power in hp} = \frac{F \times s}{550t}$$

where:

$$F = \text{force in lb}$$
$$s = \text{distance in ft}$$
$$t = \text{time in seconds}$$

or

$$\text{hp} = \frac{F \times s}{33{,}000t}$$

where $t =$ time in minutes.

Other ways of writing Equation 14-6 would be as follows:

$$P = \frac{F \times s}{t}, \text{ but } v = \frac{s}{t}$$

therefore:

$$P = F \times v$$

For rotational motion, recall that for angular work:

$$U = \text{torque} \times \theta \qquad \text{(Equation 14-5)}$$
$$P = \frac{U}{t} = \frac{\text{torque} \times \theta}{t}, \text{ but } \omega = \frac{\theta}{t}$$

therefore:

$$P = \text{torque} \times \omega \qquad \text{(14-7)}$$

where:

$$\text{torque is in N} \cdot \text{m}$$
$$\omega = \text{angular velocity in rad/s}$$
$$P = \text{power in watts (W)}$$

and, in the English system:

> torque is in lb-ft
> $\omega =$ angular velocity in rad/sec
> P = power in ft-lb/sec

We can convert between English and SI by using the conversion factor:

$$1 \text{ horsepower (hp)} = 0.746 \text{ kW}$$

Example 14-18

An average force of 300 N is applied over a distance of 50 m. If the time required is 2 min, determine (a) the work, and (b) the power.

$$\text{work} = F \times s$$
$$= 300 \times 50$$
$$= 15,000 \text{ J}$$
$$\underline{\text{work} = 15 \text{ kJ}}$$

$$\text{power} = \frac{\text{work}}{\text{time}}$$
$$= \frac{15,000}{60 \times 2}$$
$$\underline{\text{power} = 125 \text{ W}}$$

Example 14-19

Determine the horsepower required to provide a force of 400 lb for a distance of 8 ft in a time of 5 seconds.

$$\text{hp} = \frac{F \times s}{550t}$$
$$= \frac{400 \times 8}{550 \times 5}$$
$$\underline{\text{hp} = 1.16 \text{ hp}}$$

Various machines are capable of receiving power and converting it to a more useful form. An electric motor receives input energy of kilowatts and gives an output of horsepower. The efficiency at which a machine can transmit or convert this energy is described by the ratio of power output to power input. Expressed as a percentage:

$$\text{efficiency (in percent)} = \frac{\text{power output}}{\text{power input}} \times 100$$

An interesting example of this overall *efficiency* is that of a car with an internal combustion gasoline engine. If the total heat input of the gasoline were converted at 100 percent efficiency to work causing motion of the car (with no friction or wind resistance), then it would yield an astonishing mileage of approximately 450 miles per gallon (159 km/l).

Example 14-20

The starting motor on a turbine applies a constant torque of 60 lb-ft while turning the turbine 250 revolutions in 15 sec. Calculate the horsepower required.

$$U = \text{torque} \times \theta \qquad \text{(Equation 14-5)}$$

$$\text{power} = \frac{U}{t} = \frac{\text{torque} \times \theta}{t}$$

since

$$1 \text{ hp} = 550 \text{ ft-lb/sec}$$

$$P = \frac{\text{torque} \times \theta}{550t}$$

$$= \frac{60(250 \times 2\pi)}{550 \times 15}$$

$$P = 11.4 \text{ hp}$$

Example 14-21

What is the efficiency of an electric motor that supplies 8 hp while using 7.1 kW of electricity?

$$\text{efficiency} = \frac{\text{output}}{\text{input}} \times 100$$

$$= \frac{8 \times 0.746 \text{ kW/hp}}{7.1 \text{ kW}} \times 100$$

$$\text{efficiency} = 84 \text{ percent}$$

Note that, if we had been using the SI metric system, we would have had the motor supplying 5.96 kW of power while using 7.1 kW of electricity. Efficiency would have been easily calculated without any cumbersome conversion; that is:

$$\text{efficiency} = \frac{5.96}{7.1} \times 100$$

$$\text{efficiency} = 84 \text{ percent}$$

Example 14-22

A cutting tool on a lathe applies a tangential force of 3.5 kN when it is machining a bar 150 mm in diameter. If the lathe is turning at 100 rpm, what power must be supplied to the bar?

$$P = \frac{\text{torque} \times \theta}{t}$$

$$= \frac{\left(3500 \times \dfrac{0.15}{2}\right)(100 \times 2\pi)}{60}$$

$$P = 2.75 \text{ kW}$$

PROBLEMS

14-1 Determine the work done on a 100-lb body that is moved 6 ft horizontally by a 20-lb horizontal force.

14-2 Determine the work done in lifting a 100-lb body to a height of 6 ft.

14-3 A man pushes a loaded cart 15 m horizontally by pushing with a horizontal force of 60 N. He then unloads 20 bags, each with a mass of 40 kg, from the cart. Each bag is lifted to a height of 0.8 m. Determine the total work done.

14-4 Slider A in Figure P14-4 is lifted to a vertical height of 5 m by the rope shown. The rope remains at the same angle throughout the motion. Determine the work done on A by the rope.

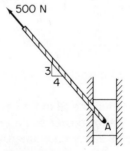

Figure P14-4

14-5 Box A is moved 15 ft to the right by the force shown in Figure P14-5. Determine the work done on the box by the 200-lb force.

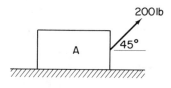

Figure P14-5

14-6 A construction worker must apply the horizontal and vertical forces shown in Figure P14-6 while pushing the loaded wheelbarrow 6.5 m up the slope. Determine the work done.

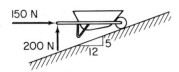

Figure P14-6

14-7 Determine the work done on block A by the forces shown in Figure P14-7 when A moves 10 ft to the right.

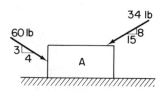

Figure P14-7

14-8 Block A in Figure P14-8 has a mass of 15 kg and is pushed 3.4 m up the slope by a force, $P = 130$ N. Determine the work done on the block by (a) force P, and (b) the friction force.

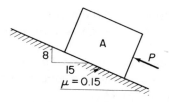

Figure P14-8

14-9   Cart A is pushed 6 m along the slope by the 400 N force shown in Figure P14-9. Determine the work done on the car by the 400 N force.

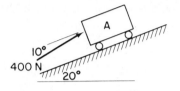

Figure P14-9

14-10   Cart A weighs 390 lb, and a force of 260 lb is applied to it as shown in Figure P14-10. Determine the work done on the cart when it moves 10 ft along the slope.

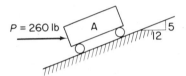

Figure P14-10

14-11   Neglect rolling resistance and determine the minimum amount of work required to push cart A, in Problem 14-10, 5 ft up the slope.

14-12   A 200-kg block is slowly pushed a distance of 80 m up a slope that makes an angle of 30° with the horizontal. Determine the work done by a horizontally applied force *P* if the coefficient of friction is 0.3.

14-13   A 40-kg block slides 5 m down a 60° slope. If the coefficient of friction is 0.4, determine the work done on the block by the friction force.

14-14   What force is required to stretch a spring that has a spring constant of 0.4 lb/in. 3 in. beyond its free length?

14-15   A spring is compressed 4 in. from its free length by a force of 80 lb. Determine the spring constant and the work required.

14-16   A spring is compressed 200 mm from its free length by a force of 500 N. Determine the spring constant and the work required.

14-17   A spring is initially compressed 50 mm by a force, *P*. If *P* is increased by 250 N thereby causing a total spring compression of 150 mm, what is the spring constant?

14-18 Determine the work required to stretch a spring 6 in. from its free length if it has a spring constant of 2 lb/in.

14-19 A shock absorber spring, initially at its free length, absorbs 1800 J of energy while deflecting 420 mm. Determine the spring constant.

14-20 A spring that has a spring constant of 25 lb/in. slowly absorbs 200 in·lb of energy while deflecting from its free length. How much is it compressed?

14-21 The spring shown in Figure P14-21 has a free length of 0.5 m and a spring constant of 1.2 kN/m. Determine the work required to move the lever to the vertical position.

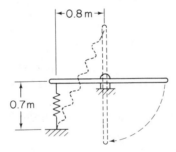

Figure P14-21

14-22 The spring shown in Figure P14-22 has a free length of 160 mm and a spring constant of 800 N/m. Neglect all friction and determine how much work the spring does in lifting block A 50 mm?

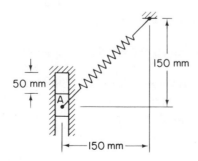

Figure P14-22

14-23 A spring that is initially compressed 3 in. by a force of 90 lb is compressed 2 in. farther. Determine the work required to compress the spring by the additional 2 in.

14-24 The spring in Figure P14-24 is stretched 0.1 m so that it is in the position shown. It has a spring constant of 3 kN/m. Determine the work required to rotate arm AB 90° clockwise.

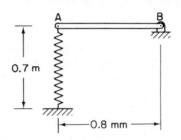

Figure P14-24

14-25 A mass of 26 kg is suspended from a slider, as shown in Figure P14-25, and moves from point A to point B. Neglect the mass of the slider and assume a coefficient of friction of 0.2. Determine the work done on the slider by (a) the 26-kg mass and (b) the friction forces.

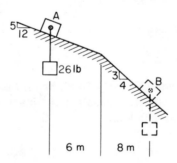

Figure P14-25

14-26 Weight A in Figure P14-26 is winched to the right by a rope tension that varies linearly from 30 lb to 40 lb. Determine the work done by the rope on weight A when it is pulled 8.5 ft to the right as shown.

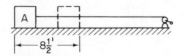

Figure P14-26

14-27   Determine the kinetic energy of a truck weighing 40 kN and travelling at 25 m/s. How high would the equivalent amount of energy lift the truck?

14-28   A carton being stacked in a warehouse falls 20 ft. Determine the velocity at which it lands.

14-29   A construction crane drops a mass of 50 kg from a height of 4 m. Determine the kinetic energy of the mass as it strikes the ground and the velocity at which it strikes the ground.

14-30   A 32.2-lb object is dropped from a 100-ft building. Determine the velocity and kinetic energy at (a) the 50-ft level, and (b) the 25-ft level above the ground.

14-31   An object has 800 J of kinetic energy when it is moving at 1 m/s. Determine its kinetic energy when its velocity is (a) 2 m/s, and (b) 3 m/s.

14-32   Determine the velocity of weight A in Figure P13-9 after it has moved 2 m from rest.

14-33   Determine the velocity of weight B in Figure P13-10 after it has moved 3.5 m from rest.

14-34   Use the work-energy method to solve Problem 13-17.

14-35   Use the work-energy method to solve Problem 13-15 if the force $P = 100$ lb moves the block 40.25 ft.

14-36   A 70-kg man riding a 120-kg motorcycle starts from rest and accelerates at 5 m/s² for 5 s. Determine the total kinetic energy at $t=5$ s.

14-37   If mass A in Figure P13-20 is 30 kg and starts from rest, how far has it moved when its velocity is 5 m/s?

14-38   Use the work-energy method to solve Problem 13-26.

14-39   A 48-tonne ferry strikes a dock at 3 km/h. What average force does it exert on the dock if it is brought to rest over a distance of 0.2 m?

14-40   Use the work-energy method to solve Problem 13-25.

14-41   Door A, in Figure P14-41, weighs 50 lb. The friction and inertia of the pulleys and rollers can be neglected. If counterweight B weighs 15 lb, determine the velocity of the door if, after starting from rest, it moves 5 ft to the right.

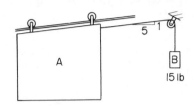

Figure P14-41

14-42   The spring in Figure P14-42 has a spring constant of 1.5 kN/m and a free length of 0.6 m. If the spring starts from rest at the position shown, determine the distance that A moves if it is (a) dropped, and (b) lowered slowly.

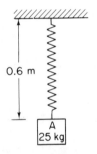

Figure P14-42

14-43   If the spring in Problem 14-42 has a 25-kg mass dropped from the position shown in Figure P14-43, determine the distance that the mass drops.

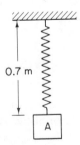

Figure P14-43

14-44   If the spring in Problem 14-42 has a 25-kg mass released from the position shown in Figure P14-44, determine the maximum distance upward that the mass can move.

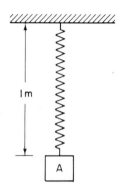

Figure P14-44

14-45   A steel ball with sufficient velocity will roll through the hoop as shown in Figure P14-45. Determine the velocity of the ball at A if, when it reaches point B, it is on the verge of dropping away from the track.

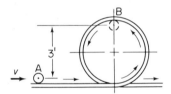

Figure P14-45

14-46   Cart A, weighing 390 lbs, has an initial velocity of 15 ft/sec at the position shown in Figure P14-46. If the cart has a constant rolling resistance of 10 lb, determine the spring deflection and the location of the cart with respect to its initial position when it comes to rest on the horizontal surface. The spring constant is 300 lb/in., and the rolling resistance can be neglected for the distance over which the spring is compressed.

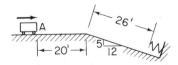

Figure P14-46

14-47 The 10-kg block A in Figure P14-47 is pulled down to the left so that the spring is compressed 80 mm. Determine the velocity of block A when it reaches point B.

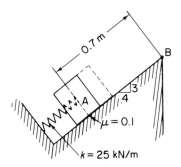

Figure P14-47

14-48 Collar A in Figure P14-48 weighs 10 lb and slides on a frictionless rod. The spring has a free length of 7 in. If A is released from the position shown, determine the velocity of A at point B.

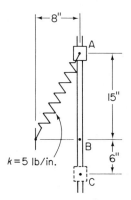

Figure P14-48

14-49 Use the data in Problem 14-48 to determine the velocity of A when it reaches point C.

14-50 A 5-kg gear has a radius of gyration of 200 mm. Determine its kinetic energy when it is rotated at 90 rpm.

14-51 The front wheel of a car weighs 25 lb and has a radius of gyration of 0.5 ft. While being dynamically balanced after the installation of a tire, the wheel is rotated at 80 rpm. Determine its kinetic energy.

14-52 Determine the kinetic energy of a 2.5-m slender rod, with a mass of 10 kg, when it is rotated at 20 rpm about A in Figure P14-52.

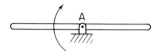

Figure P14-52

14-53 Determine the kinetic energy of the rod in Problem 14-52 when it is rotated about an axis through one end.

14-54 A shaft, 5 ft long and 4 in. in diameter, weighs 220 lb. A 322-lb rotor with a radius of gyration of 15 in. is mounted on the shaft. For a speed of 1500 rpm, determine the kinetic energy of (a) the shaft; (b) the rotor; and (c) the shaft and rotor combined.

14-55 A nut is tightened during its last half turn by an average torque of 20 N·m. Determine the work done.

14-56 A shaft coupling is rated for a torque of 100 N·m at 300 rpm. Determine the energy that it transmits in one minute.

14-57 The flywheel on a punch press turns two revolutions while supplying 3000 ft-lb of energy to punch a hole. Determine the average torque supplied during this time.

14-58 Wheel A in Figure P14-58 weighs 200 lb and has a radius of gyration of 2 ft. If the system is initially at rest, determine the angular velocity of A after B has dropped 8 ft.

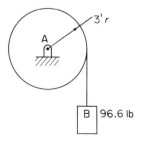

Figure P14-58

14-59   Masses A and B in Figure P14-59 are fastened together by a belt over pulley D. (Assume no slipping of the belt.) The mass moment of inertia of pulley D is 15 kg·m². How far does mass B drop before reaching a velocity of 2 m/s?

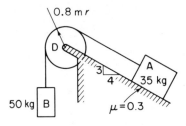

Figure P14-59

14-60   Drum A in Figure P14-60 has a mass moment of inertia of 65 kg·m². If B has a velocity of 1.5 m/s downward, determine the force P necessary to brake drum A to a stop in one revolution.

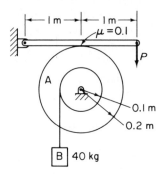

Figure P14-60

14-61   Double-pulley D in Figure P14-61 has a mass moment of inertia of 150 ft-lb-sec². If the system is initially at rest, determine the velocity of A just before B strikes the ground.

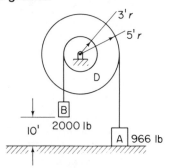

Figure P14-61

14-62 The spring in Figure P14-62 is at its free length, and the system is at rest when weight B is dropped. Determine the maximum distance that B will drop.

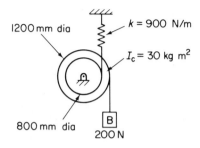

1200 mm dia $k = 900$ N/m

$I_c = 30$ kg m^2

800 mm dia   B
                    200 N

Figure P14-62

14-63 Pulley A in Figure P14-63 weighs 322 lb and has a mass moment of inertia of $I_c = 50$ ft-lb-sec^2. The system is at rest and the spring is at its free length for the position shown. Determine the maximum distance that B will drop when released.

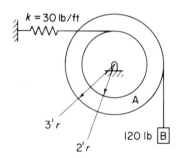

$k = 30$ lb/ft

3' r

2' r

A

120 lb B

Figure P14-63

14-64 The wheel shown in Figure P14-64 has an initial velocity and, while coming to rest, turns 200 revolutions at constant deceleration. If the average friction force acting on piston B is 10 N, determine the initial speed at the wheel. (Neglect all other friction and the mass of the piston and member AB).

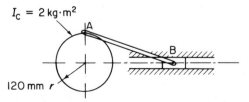

$I_c = 2$ kg·m^2

A

B

120 mm r

Figure P14-64

14-65 Cylinder A in Figure P14-65 weighs 32.2 lb. The spring is at its free length and has a spring constant of 10 lb/ft. If the system is initially at rest, determine the velocity of B after it has dropped 6 ft.

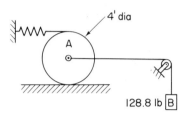

Figure P14-65

14-66 Use the work-energy method to solve Problem 13-45.

14-67 A log rolls without slipping down a 45° slope that is 100 ft long. The log weighs 800 lb and has an average radius of 9 in. When the log is at the bottom of the slope, determine (a) its linear kinetic energy and (b) its rotational kinetic energy.

14-68 Use the work-energy method to solve Problem 13-53 if B has a velocity of 7.62 m/s at $t = 20$ s.

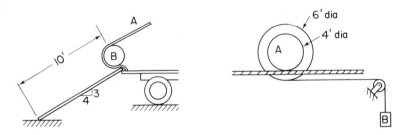

Figure P14-69   **Figure P14-70**

14-69 A 50-lb drum, 3 ft in diameter, is lowered down planking by means of a rope as shown in Figure P14-69. If the rope starts to slip in the operator's hands at A and he then allows the rope to slip freely, what will be the velocity of the rope through his hands when the drum reaches the bottom of the plank?

14-70 Block B in Figure P14-70 weighs 20 lb, and A weighs 32.2 lb. If they start at rest, determine the angular velocity of A when B has dropped 10 ft. (Assume no slippage of A.) The mass moment of inertia of A about its center is 2.8 ft-lb-sec².

14-71 Use the work-energy method to solve Problem 13-56.

14-72 The combined wheel and hub in Figure P14-72 has a mass of 25 kg and a mass moment of inertia of 27 kg·m². Mass of block B is 12 kg. If the center of A has a velocity of 1.5 m/s to the left at the instant shown and rolls without slipping, determine the distance that B will move before bringing the system to rest.

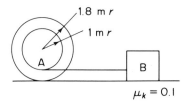

Figure P14-72

14-73 A belt wound around cylinder B in Figure P14-73 supports cylinder A. If the system is initially at rest, determine the angular velocity of A when it has dropped 1.5 m.

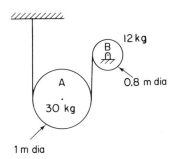

Figure P14-73

14-74 A 180-lb man can climb a 50-ft vertical ladder in 30 sec. Determine the power in hp that the man requires for this.

14-75 Convert 50 kW to ft-lb/sec.

14-76 A winch 2 ft in diameter lifts a 3000-lb weight at 10 ft/sec. What horsepower and torque must be supplied to the winch drum?

14-77 A hydraulic hoist is to lift 3 tonnes a height of 4 m in 15 s. To what power requirement is this equivalent?

14-78   A revolving restaurant has a 100-ft diameter, turns at one revolution per hour, and is driven by a 1-hp motor. Assume that two-thirds of this power is lost due to inefficiency and gearing losses and convert the remaining available power into a tangential force that would give the same speed of rotation. At what velocity is the outer edge travelling?

14-79   A turbojet flying at 450 mph has a 5000-lb thrust. What horsepower does this represent?

14-80   What is the power output of an 85% efficient electric motor that uses 200 W of electrical power?

14-81   What would be the horsepower output of a 107-kW motor that is 90% efficient?

14-82   A man weighing 700 N climbs to a vertical height of 15 m in 30 s. Determine the power developed.

14-83   An electric traction motor provides a torque of 400 N·m while rotating through 60 revolutions in 5 s. If the motor is 90% efficient, determine the electrical power input.

14-84   The centrifugal discharge bucket elevator in Figure P14-84 must lift 40 tons per hour to a vertical height of 80 ft. If the drive efficiency is 75%, determine the required output horsepower of the motor that drives the elevator.

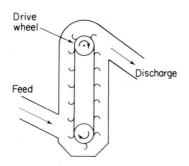

Figure P14-84

14-85   Driving a 2000-ft loaded horizontal conveyor at 4 ft/sec requires 35 hp. What horsepower would be required if the conveyor is now raised at one end giving a gradient of 1-ft rise in 60 ft of horizontal run? Each foot of conveyor carries a load of 20 lb.

IMPULSE AND MOMENTUM

15-1   LINEAR IMPULSE AND MOMENTUM

An object that is subjected to a force for a very short time is said to have received an *impulse*; impulse is equal to force times time. The movement or change in velocity of the object depends on the mass of the object and is described by the term *momentum*, which is equal to mass times velocity.

The object receiving a sudden blow has in effect had a variable force applied. Considering the average value of this force, we can write:

$$F = ma$$

where

$$a = \frac{v}{t}$$

Therefore:

$$F = \frac{mv}{t}$$

$$Ft = mv \qquad\qquad (15\text{-}1)$$

where *Ft* is *impulse* and *mv* is *momentum*. The change in momentum of an object is equal to the impulse applied to the object.

In the SI system:

$$F = \text{force in N}$$
$$t = \text{time in s}$$
$$m = \text{mass in kg}$$
$$v = \text{velocity in m/s}$$

Substituting these units into Equation 15-1, we have:

$$N \times s = kg \times m/s$$

but

$$1\ N = 1\ kg \times 1\ m/s^2$$

or

$$1\ kg = 1\ Ns^2/m$$

Substituting this into our impulse = momentum equation, we have:

$$N \times s = \frac{N\ s^2}{m} \times \frac{m}{s}$$
$$N \cdot s = N \cdot s$$

For linear motion, the units of both impulse and momentum are newton-seconds.

The common units in the English system are:

$$f = \text{force in lb}$$
$$t = \text{time in sec}$$
$$m = \text{mass in slugs or } \frac{\text{lb-sec}^2}{\text{ft}}$$
$$v = \text{velocity in ft/sec}$$

The units of impulse and momentum can be found by substitution of these units in Equation 15-1 (*Ft* = *mv*).

$$lb \times sec = lb\text{-}sec^2/ft \times ft/sec$$
$$lb\text{-}sec = lb\text{-}sec$$

To handle situations in which the initial velocity is not zero, Equation 15-1 can be written as:

$$F\Delta t = m\Delta v$$

This equation is the basis of the impulse = momentum method referred to earlier. Both impulse and momentum are vector quantities; therefore, direction must be taken into account when this formula is used.

Although our initial description of impulse was that of a force applied for a short period of time, the length of time can be longer as the following example will show.

Example 15-1

If there is a constant towing force of 1.3 kN, how long would it take a tow truck to tow a 1500-kg car from rest to 90 km/h? (Assume a rolling resistance of 200 N.)

$$Ft = mv$$

$$(1300 - 200)t = 1500 \times \frac{90,000}{3600}$$

$$\underline{t = 34.1 \text{ s}}$$

Example 15-2

A 1.6-oz golf ball is struck by a club so that a force of 18 lb is applied for 0.04 sec. What is the velocity of the ball due to this impulse?

$$Ft = mv$$

$$18 \times 0.04 = \frac{1.6}{16} \times \frac{1}{32.2} v$$

$$\underline{v = 232 \text{ ft/sec}}$$

15-2   ANGULAR IMPULSE AND MOMENTUM

Rotating objects are also subject to impulse. The change in motion is not simply due to a force but rather to a force acting at some radius. *Angular impulse* is therefore due to a torque. From Section 13-3, we can write:

$$\text{(torque) } T = I_c \alpha \text{ where } \alpha = \frac{\omega}{t}$$

therefore:

$$T = \frac{I_c \omega}{t}$$

$$Tt = I_c \omega \qquad \qquad (15\text{-}2)$$

where, in the SI system:

   T = torque in N·m
   t = time in s
   I_c = mass moment of inertia about the center of mass in kg·m^2
   ω = angular velocity in rad/s

In the English system, the units are:

   T = torque in lb-ft
   t = time in sec
   I_c = mass moment of inertia about the center of mass in ft-lb-sec^2
   ω = angular velocity in rad/sec

To handle all cases of speed change in a given time period—not just those in which the object is initially at rest—Equation 15-2 could also be written as:

$$T\Delta t = I_c \Delta \omega$$

This formula could be applied to pure rotation, such as occurs in a flywheel. When we come to plane motion such as with a rolling wheel, linear impulse and linear momentum must also be considered.

Example 15-3

A torque of 30 lb-ft is applied to a turbine wheel that has a mass moment of inertia of 60 ft-lb-sec^2 about its center of rotation. Neglect bearing friction and determine the angular velocity that the turbine acquires in 3 sec from an initial state of rest.

$$Tt = I_c \omega$$

$$30 \times 3 = 60\omega$$

$$\underline{\omega = 1.5 \text{ rad/sec}}$$

Example 15-4

A drum with a mass of 120 kg and 0.4 m in diameter rolls down a plane inclined 15° to the horizontal. If the drum's initial speed is 0.8 m/s and no slippage occurs, determine its speed 10 s later.

The mass moment of inertia is:

$$I_c = \frac{1}{2} mr^2$$

$$= \frac{1}{2} \times 120 \times (0.2)^2$$

$$I_c = 2.4 \text{ kg} \cdot \text{m}^2$$

The initial velocities are:

$$v_0 = 0.8 \text{ m/s}$$

$$\omega_0 = \frac{v_0}{r} = \frac{0.8}{0.2} = 4 \text{ rad/s}$$

The final velocities are v and:

$$\omega = \frac{v}{0.2}$$

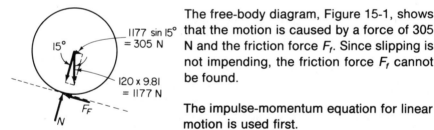

The free-body diagram, Figure 15-1, shows that the motion is caused by a force of 305 N and the friction force F_f. Since slipping is not impending, the friction force F_f cannot be found.

Figure 15-1

The impulse-momentum equation for linear motion is used first.

$$F\Delta t = m\Delta v$$
$$(305 - F_f)10 = 120(v - 0.8)$$
$$F_f = 315 - 12v \qquad (1)$$

Writing the impulse-momentum equation for angular motion, we have:

$$T\Delta t = I_c \Delta \omega$$
$$(F_f \times 0.2)10 = (2.4)\left(\frac{v}{0.2} - 4\right)$$
$$2F_f = 12v - 9.6 \qquad (15\text{-}4)$$

Substituting Equation 1 into Equation 2, we get:

$$2(315 - 12v) = 12v - 9.6$$
$$\underline{v = 17.8 \text{ m/s}}$$

15-3   CONSERVATION OF MOMENTUM

A moving object may transfer or lose some of its momentum to another object. The momentum is transferred by means of impulse. Consider a fast-moving object that strikes and becomes attached to a slow-moving object travelling in the same direction. The fast-moving object exerts a force or impulse on the slow-moving object, thereby speeding it up. The second object at that instant exerts an equal and opposite reaction or impulse on the first object, thereby slowing it down. This action is summarized in Newton's third law which states, "For every action there is an equal and opposite reaction."

While each object has experienced a change in velocity and therefore a change in momentum, the total momentum of the system remains the same. This is of course known as *conservation of momentum* and applies to both linear momentum and angular momentum.

In this section, we will consider only *inelastic* types of impact, i.e., where the objects become attached to each other on impact.

Note two other points: we are talking about vector quantities here; and there are no external forces being applied to the system.

Referring to Figure 15-2, we see that conservation of linear momentum can be written as:

$$\text{initial momentum} = \text{final momentum}$$
$$(m_A v_A)_1 + (m_B v_B)_1 = (m_A v_A)_2 + (m_B v_B)_2$$

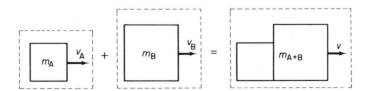

Figure 15-2

since

$$v=(v_A)_2=(v_B)_2$$
$$m_A v_A + m_B v_B = (m_{A+B})v$$

Angular momentum can be shown in a similar manner (Figure 15-3).

initial momentum = final momentum
$$(I_A\omega_A)_1 + (I_B\omega_B)_1 = (I_A\omega_A)_2 + (I_B\omega_B)_2$$

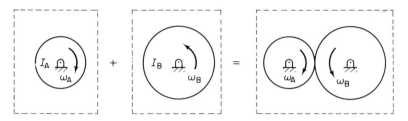

Figure 15-3

In Figure 15-4, there is a common angular velocity, ω; therefore:

$$(I_A\omega_A)_1 + (I_B\omega_B)_1 = (I_A\omega)_2 + (I_B\omega)_2$$
$$(I_A\omega_A) + (I_B\omega_B) = (I_A + I_B)\omega$$

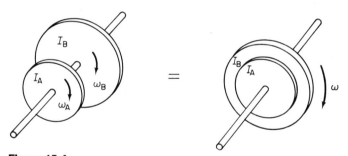

Figure 15-4

In applying any of these equations, we must give velocity a negative or positive sign. Choose the direction for positive velocity and be consistent throughout. If you are in doubt about the direction of a velocity, assume the positive direction, and the sign of the result will then agree with the previously chosen sign convention.

It is not necessary to have two objects in order to apply the principle of conservation of momentum. If a single rotating object with a given initial momentum ($I\omega_O$) had its configuration changed so that its moment of inertia were reduced, then there must have been a corresponding increase in angular velocity.

Example 15-5

A whirling figure skater with her arms extended has an angular velocity of 12 rad/s. Determine her angular velocity when, by drawing her arms close to her body, she reduces her moment of inertia by 40%.

$$\text{initial momentum} = \text{final momentum}$$
$$I_0\omega_0 = I\omega$$
$$I_0 \times 12 = 0.6 I_0 \omega$$
$$\underline{\omega = 20 \text{ rad/s}}$$

Example 15-6

A belt-driven pulley is used to drive a machine that has a constant torque requirement of 60 N·m. The pulley has a mass moment of inertia, $I_C = 0.7$ kg·m², a diameter of 0.4 m, and a mass of 25 kg. The pulley was initially turning at 300 rpm, but the belt tensions were increased to 500 N and 150 N. Determine the new pulley speed after 3 s at the new tensions.

$$\omega_0 = \frac{300 \times 2\pi}{60} = 31.4 \text{ rad/s}$$

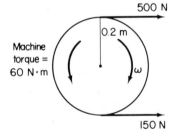

Machine torque = 60 N·m

The net torque on the pulley is input torque − output torque (Figure 15-5).

$$\text{net torque} = \left[(500 - 150)0.2\right] - 60$$
$$\text{net torque} = 10 \text{ N·m}$$
$$\text{impulse change} = \text{momentum change}$$
$$\Delta T t = \Delta I \omega$$

Figure 15-5

or

$$\text{net torque} \times t = I \times \Delta\omega$$
$$10 \times 3 = 0.7(\omega - 31.4)$$
$$\omega = 74.3 \text{ rad/s}$$
$$\underline{\omega = 709 \text{ rpm}}$$

Example 15-7

A 4000-lb car (A) and a 3000-lb car (B) collide on glare ice and remain together. At the time of impact, car A was travelling east at 50 mph and car B was travelling north at 30 mph. Determine their resulting velocity and direction.

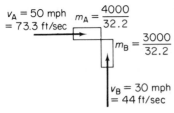

$v_A = 50$ mph
$= 73.3$ ft/sec

$m_A = \dfrac{4000}{32.2}$

$m_B = \dfrac{3000}{32.2}$

$v_B = 30$ mph
$= 44$ ft/sec

Figure 15-6

The mass and velocity of each car are shown in Figure 15-6. Initial momentum equals final momentum, but the initial momentum values must be added vectorially (Figure 15-7). The direction of the final velocity will be the same as that of the final momentum.

$$m_A v_A + m_B v_B = (m_A + m_B) v$$

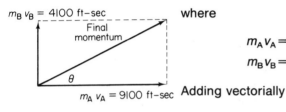

$m_B v_B = 4100$ ft–sec

Final momentum

θ

$m_A v_A = 9100$ ft–sec

Figure 15-7

where

$$m_A v_A = 9100 \text{ ft-sec} \rightarrow$$
$$m_B v_B = 4100 \text{ ft-sec} \uparrow$$

Adding vectorially

$$(m_A + m_B)v = \sqrt{(9100)^2 + (4100)^2}$$
$$(m_A + m_B)v = 9980$$
$$v = \dfrac{9980}{\dfrac{4000 + 3000}{32.2}}$$
$$\underline{v = 45.8 \text{ ft/sec} \nearrow 24.2°}$$

$$\tan\theta = \dfrac{4100}{9100}$$
$$\underline{\theta = 24.2°}$$

Example 15-8

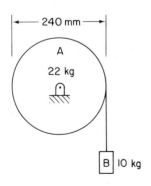

—— 240 mm ——

A

22 kg

B 10 kg

Figure 15-8

Cylinder A in Figure 15-8 has a radius of gyration of 180 mm. The system is initially at rest and has no bearing friction. Determine the angular velocity of A, the angular acceleration of A, and the tension in the rope at $t = 3$ s.

This example is similar to Example 13-9, which was solved by the force-inertia method.

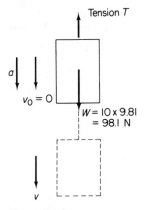

Figure 15-9

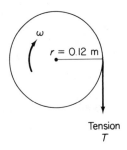

Figure 15-10

First, we will apply the impulse-momentum equation to block B (Figure 15-9).

Net impulse change = momentum change

$$\Delta Ft = \Delta mv$$

$$(W - T)t = \frac{W}{g}(v - 0)$$

$$(98.1 - T)3 = 10v$$

$$T = 98.1 - 3.33v$$

but

$$v = r\omega = 0.12\omega$$

$$T = 98.1 - 0.4\omega \qquad (1)$$

Now apply the angular impulse-momentum equation to cylinder A (Figure 15-10).

$$(\text{torque}) \times \Delta t = I_c\omega$$

where

$$I_c = k^2m$$

$$= (0.18)^2 \times 22$$

$$I_c = 0.713 \text{ kg} \cdot \text{m}^2$$

$$(T \times 0.12)3 = 0.713 \times \omega$$

$$0.36T = 0.713\omega \qquad (2)$$

Substituting Equation 1 into Equation 2, we get:

$$0.36(98.1 - 0.4\omega) = 0.713\omega$$

$$\underline{\omega = 41.2 \text{ rad/s}}$$

Substituting this into Equation 1, we get:

$$\underline{T(\text{tension}) = 81.7 \text{ N}}$$

For the cylinder:

$$\alpha = \frac{\Delta\omega}{t}$$

$$= \frac{41.2 - 0}{3}$$

$$\underline{\alpha = 13.7 \text{ rad/s}^2}$$

PROBLEMS

15-1 Determine the momentum of a 90-kg mass moving with a velocity of 60 m/s.

15-2 Determine the momentum of a 200-lb man running at 10 mph.

15-3 A 2-lb object on a smooth horizontal surface is subject to a horizontal force of 1 lb for 20 sec. Determine the velocity at $t=20$ sec.

15-4 For how long does a 200-N force have to be applied to increase the velocity of a 30-kg mass from 80 to 150 m/s?

15-5 A soccer player kicks a 0.3-lb ball with an average force of 5 lb giving it a velocity of 75 ft/sec. For how long was the force applied?

15-6 For how long does a 200-N force have to be applied to change the momentum of an object by 800 N·s?

15-7 A 1300-kg stationary car whose brakes are not applied is subjected to a force of 5 kN for 2 s when struck in the rear by another car. What velocity is imparted to the car? If the car that was struck had weighed 50% less, what would have been its velocity? Assume that the 5-kN force is applied for the same length of time.

15-8 A 4000-lb car is to be towed from rest by means of a rope. A force of 800 lb is applied for 3 sec before the rope breaks. Determine the velocity of the car.

15-9 The force of P is applied as shown in Figure P15-9 for 5 s. Determine the velocity of the block at $t=5$ s.

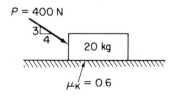

Figure P15-9

15-10 How long will it take block A in Figure P15-10 to come to rest? ($\mu k = 0.6$.)

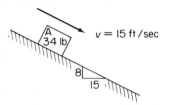

$v = 15\ \text{ft/sec}$

Figure P15-10

15-11 Solve Problem 15-10 if the mass of A is 80 kg and has an initial velocity of 4 m/s.

15-12 Arrestor barriers are used to stop aircraft that are unable to stop due to various emergency conditions (Figure P15-12). Vertical nylon ropes are attached to two horizontal cables. When an aircraft runs into the barrier, the cables unroll from winding drums at high speed; the initial breaking power is slight. Maximum braking is achieved when the drums have reached maximum speed. What is the average force required to stop a 35-ton aircraft in 6 sec when it strikes the barrier at 185 mph?

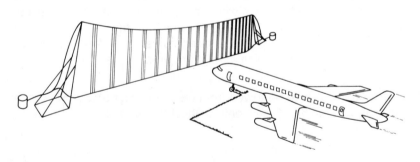

Figure P15-12

15-13 Solve Problem 13-15 by the impulse-momentum method.

15-14 Solve Problem 13-24 by the impulse-momentum method.

15-15 A pulp and paper machine has a roller that has a mass of 400 kg and a mass moment of inertia about its center of rotation of 30 kg·m². How long will it take a torque of 15 N·m to accelerate the machine from 0.8 to 2.3 revolutions per second?

15-16   A pulley weighs 200 lb and has a mass moment of inertia of 4 ft-lb-sec². How long will it take a torque of 15 lb-in. to accelerate the pulley from 50 to 150 rpm?

15-17   A flywheel ($I_C = 25$ kg·m²) on a baler is free-wheeling at 200 rpm immediately after its shear pin fails. What tangential friction force must be applied to a flywheel 0.44 m in radius to stop it in 3 s?

15-18   Solve Problem 13-29 by the impulse-momentum method.

15-19   Inertia welding of two shafts end-to-end is accomplished by rotating one shaft and bringing it against the end of the other fixed shaft (shafts A and B in Figure P15-19). When sufficient heat is generated due to friction, the driving fixture on shaft A releases and shaft A comes to rest. It takes 2 s to weld two 200 mm diameter, 2-m long shafts when shaft A is initially rotating at 800 rpm. Determine the average torque contributed by shaft A alone. The density of the steel in the shafts is 0.00783 kg/cm³.

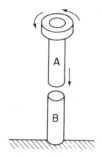

Figure P15-19

15-20   A helicopter rotor has a mass moment of inertia of 6500 kg·m². If it is initially at rest, determine its speed after a torque of 7000 N·m is applied for 45 s.

15-21   Solve Problem 13-30 by the impulse-momentum method.

15-22   An 80-mm square, 0.3-m long wood block is mounted longitudinally in a lathe and turned at 200 rpm. It is then machined down to an 80-mm diameter. Determine the difference in angular momentum. Assume that the wood weighs 0.00084 kg/cm³.

15-23 A steel shaft, 150 mm in diameter and 3.2 m long, transmits a constant torque of 300 N·m at 2400 rpm. Assume that the machine it is driving has no rotational inertia and determine how long it would take the shaft to coast to a stop if its input power were removed. The steel weighs 0.00783 kg/cm³.

15-24 Solve Problem 13-33 by the impulse-momentum method.

15-25 Cylinder A in Figure P15-25 weighing 644 lb has a radius of gyration, $k=0.5$ ft. Starting from rest and with no slipping, it reaches a velocity of 20 ft/sec in 10 sec. Determine the force, P necessary to cause this motion.

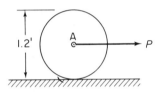

Figure P15-25

15-26 Solve Problem 15-25 if A has a mass of 250 kg, a radius of gyration of 150 mm, and a diameter of 0.4 m. It reaches a velocity of 5 m/s in 10 s.

15-27 A pipe with a 3-in. outside diameter and a 2-1/2-in. inside diameter weighs 96.6 lb. It rolls from rest down a 15° slope. Assume no slipping and determine its angular velocity 5 sec later.

15-28 A torque of 40 N·m is applied to a 100-kg armature of an electric generator for 3 s. If the speed increase is 400 rpm, determine the radius of gyration of the armature.

15-29 An 80-kg man jumps vertically downward from a bridge into a 200-kg boat that is travelling at 4 m/s. Determine the velocity of the boat after the man lands on it.

15-30 A 30-ton boxcar travelling at 20 ft/sec is coupled at the rear by a 50-ton boxcar travelling at 40 ft/sec in the same direction. Determine the final velocity of the boxcars.

15-31 A cart on casters weighs 600 N and is travelling in a straight line at 3 m/s. A mass of 20 kg is dumped onto the cart with a horizontal velocity of 4 m/s at a right angle to the cart's direction. Determine the cart's final velocity.

15-32 A can filled with sand weighs 10 lb and sits on the top of a fence post. With what velocity is the can knocked off the post when struck by a bullet weighing 0.5 oz. and travelling 2000 ft/sec? (The bullet remains embedded in the sand.)

15-33 Block A in Figure P15-33 is released from the position shown. It has a velocity of 6 m/s when it strikes and sticks to block B. If the coefficient of friction is 0.3, how long does it take the blocks to come to rest? What is the maximum velocity of B?

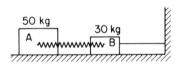

Figure P15-33

15-34 A filing cabinet drawer weighing 25 lb is slammed shut with a velocity of 8 ft/sec. If the cabinet rests on a surface with $\mu = 0.2$ and is on the verge of sliding due to the momentum of the closed drawer, for how long was the impulse applied? (The total weight of cabinet and drawers is 60 lb.)

15-35 Disk A in Figure P15-35 ($I_C = 8$ kg·m²), while rotating at 100 rpm, is lowered onto disk B ($I_C = 10.5$ kg·m²), causing the latter to rotate from rest. Assume no slipping and determine the angular velocity of B.

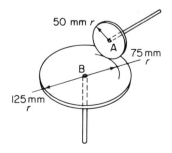

Figure P15-35

15-36 An 8 in.-diameter pulley A in Figure P15-36 is connected to a system with $I_c = 2.5$ ft-lb-sec². The system is at rest and has a friction moment of 5 lb-ft. Block B weighs 200 lb. If the system is initially at rest, at $t=4$ sec determine (a) the angular velocity of A; (b) the angular acceleration of A; and (c) the tension in the rope.

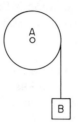

Figure P15-36

15-37 Cylinder A in Figure P15-37 rotates at 300 rpm and lifts mass B. When its source of power is removed, A and B coast to a stop in 5 s. Determine the mass of B.

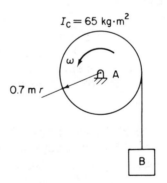

Figure P15-37

APPENDIX

The graphical solutions presented in this appendix are an alternate and possibly preferred method of solution for some problems rather than the analytical method presented earlier in the book.

The two areas in which the graphical method will be used are nonparallel, nonconcurrent force systems and truss solutions. In both cases a new system of labelling the diagram called Bow's notation will be used. In the case of trusses this notation and method yield a diagram known as Maxwell's diagram or a combined diagram.

RESULTANT OF NONPARALLEL, NONCONCURRENT FORCES

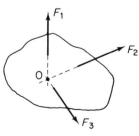

Figure A-1

In determining the resultant of concurrent forces as shown in Figure A-1, we know that the resultant will pass through point 0 and will be equal to the algebraic sum of the horizontal and vertical components of the three forces.

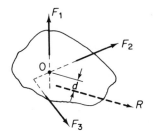

Figure A-2

If force F_3 is changed as in Figure A-2, we now have a nonparallel, nonconcurrent force system where the magnitude of the resultant will be the same but the location of the resultant will be different. This problem could be solved by the analytical or the mathematical method of using the three basic equations of

$$R_x = \Sigma F_{:x}$$
or
$$R_x = (F_1)_x + (F_2)_x + (F_3)_x$$
$$R_y = \Sigma F_y$$
$$R_y = (F_1)_y + (F_2)_y + (F_3)_y$$
$$Rd = \Sigma M$$

where moments can be taken about any point. In this case it is very convenient to take moments about the intersection of any two forces such as point 0, (Figure A-2). Since forces F_1 and F_2 have zero distances, the moment equation would be

$$Rd = F_1 d_1 + F_2 d_2 + F_3 d_3$$
$$= 0 + 0 + F_3 d_3$$
$$Rd = F_3 d_3$$

Having just considered the basic theory for the analytical method, let us now consider how these rules are applied to the graphical method.

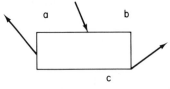

Figure A-3

To solve for the resultant of the force system shown in Figure A-3, we first label the diagram using *Bow's notation*. This is a system where lower case letters are assigned to the spaces between the vectors or their lines of action. The vector between spaces a and b is therefore vector *AB*.

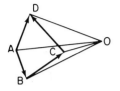

Figure A-4

Using this notation a vector polygon is drawn as shown in Figure A-4. The resultant is vector *AD* which is from the origin A to the tip of the last vector. Knowing the magnitude of resultant *AD* we must now find its location on the object of Figure A-3.

Locate a pole point 0 (Figure A-4).

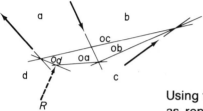

Figure A-5

Using the vectors as drawn in Figure A-3 or as reproduced in Figure A-5, perform the following construction procedure.

Draw a line ob (Figure A-5) parallel to line OB (Figure A-4) through any convenient point on the line of action of AB. Note that ob lies in space b and it intersects vector *BC*. From this intersection draw line oc parallel to OC. Continue this procedure until lines od and oa intersect.

We now have the location of the resultant (Figure A-5) and the magnitude (Figure A-4).

Example A-1

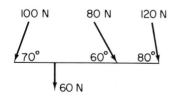

Figure A-6

Using a graphical solution, determine the magnitude and location of the resultant of the force system shown in Figure A-6.

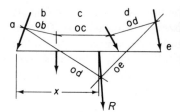

Figure A-7

Projecting the lines of action if necessary, the forces are redrawn in Figure A-7 and the spaces labelled according to Bow's notation.

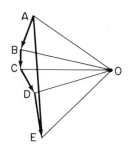

Figure A-8

Draw the vector polygon and, in upper case lettering, label as in Figure A-8.

The resultant is vector AE drawn from the origin A to the tip of the last vector. Using a scale to measure $\overline{AE}$, the magnitude of the resultant is 342N $\diagdown 85.5°$

To determine the location of the resultant, the next step necessary is to choose point 0 and draw lines to each vector tip (Figure A-8).

Lines parallel to these are transferred back to Figure A-7 in the proper sequence, starting with line ob cutting vector ab at a convenient point. Complete the sequence of oc, od, oe, and oa. The intersection of oe and oa locates the line of action of the resultant. Projecting the line of action of the resultant, the distance x (Figure A-7) is scaled to be 3.8 m.

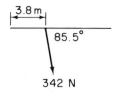

Figure A-9

The resultant of the system is therefore 342N $\diagdown 85.5°$ located 3.8 m from the left end of the member (Figure A-9).

Example A-2

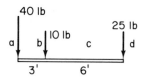

Figure A-10

Determine the magnitude and location of the resultant of the forces shown in Figure A-10.

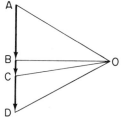

Figure A-11

While these forces are nonconcurrent as before, they are also a special case in that they are parallel. The result of this is that the vector polygon is simplified (Figure A-11). The resultant A is 75 lb$\downarrow$.

Choose point 0 and draw lines to each vector tip (Figure A-11).

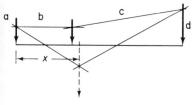

Figure A-12

Draw Figure A-12 showing the horizontal beam length to scale.

Draw line ob through a convenient point on AB and similarly draw lines oc, od, and oa. The distance measures 3.37 ft.

The resultant is therefore a force of 75 acting downward through a point 3.37 ft to the right of A.

PROBLEMS

A-1—A-9 Find and locate the resultant of the force systems shown in Figures PA-1 to PA-9.

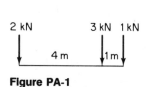

Figure PA-1

Figure PA-2

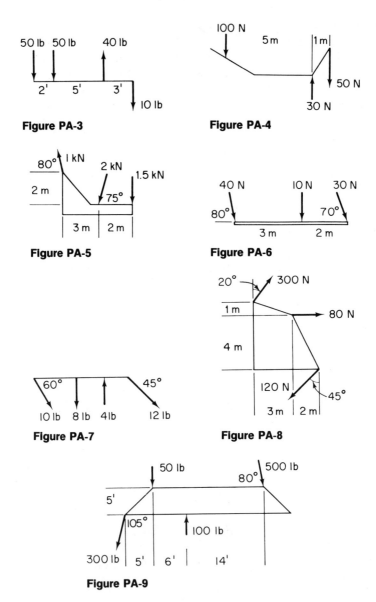

Figure PA-3

Figure PA-4

Figure PA-5

Figure PA-6

Figure PA-7

Figure PA-8

Figure PA-9

GRAPHICAL SOLUTION OF TRUSSES; MAXWELL'S DIAGRAM

When determining the load in each member of a truss by means of the analytical method of joints, free-body diagrams were drawn of each joint. The alternate graphical solution consists of drawing a vector triangle or polygon to scale of the forces at each of these joints. If these force polygons are combined or superimposed one on the other, the result is a combined diagram or *Maxwell's diagram*.

For purposes of comparison, let us rework Example 5-1 using Maxwell's diagram.

Example A-3

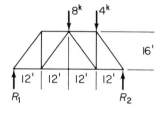

Figure A-13

Using Maxwell's diagram, determine the load in each member of the truss shown (Figure A-13).

Similar to the solution of Example 5-1, by means of a free-body diagram of the frame and moment equations we get

$$R_1 = 5 \text{ kips} \uparrow$$
$$R_2 = 7 \text{ kips} \uparrow$$

Maxwell's diagram is constructed as follows:

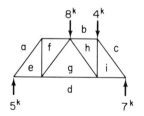

Figure A-14

STEP 1—Using Bow's notation, assign lower case letters in a clockwise direction outside the truss and then inside the truss (Figure A-14).

Figure A-15

STEP 2—Draw, to scale, a force polygon of the external forces on the truss (Figure A-15).

STEP 3—Select a joint which has one of these external forces and only two unknowns. Choosing joint aed, draw a force polygon of the forces at this point, but superimpose or add to it the force polygon of Figure A-15. Reading in a clockwise direction about the joint, we will have

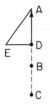

Figure A-16

forces *DA*, *AE*, and *ED*. Since *DA* was already drawn (Figure A-16), draw a line through A parallel to ae. Draw a line through D parallel to ed. The intersection of these lines gives point E (Figure A-16).

The lengths of AE and ED can be measured with a scale to get values of $AE = 6.25^k$ and $ED = 3.75^k$.

To assign tension or compression to these values, recall that we went in a clockwise direction around the joint, that is dae. Vector *DA* was drawn upward (Figure A-16). This indicates that the 5^k force was pushing upward on the joint. Vector *AE* was drawn downward to the left. This indicates member ae is pushing downward to the left on the joint; that is, ae is in compression. Following vector *ED* from left to right indicates that member ed is acting from left to right on the joint; that is, ed is in tension.

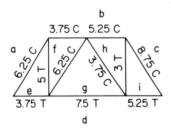

Figure A-17

STEP 4—Draw a truss sketch so that these forces and future ones can be labelled on it (Figure A-17).

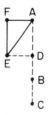

Figure A-18

STEP 5—Proceed to joint eaf and add its force polygon (Figure A-18) to the previous one. Scale the lengths and record $AF = 3.75^kC$ and $FE = 5^kT$ in Figure A-17.

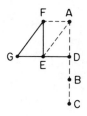

Figure A-19

STEP 6—Joint defg adds to the force polygon as in Figure A-19. Record FG$=6.25^k$C and GD$=7.5^k$T.

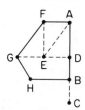

Figure A-20

STEP 7—Joint gfabh adds to the force polygon as in Figure A-20. Record BH$=5.25^k$C and HG$=3.75^k$C.

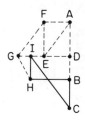

Figure A-21

STEP 8—Joint hbci adds to the force polygon as in Figure A-21. Record CI$=8.75^k$C and IH$=3^k$T. The force in DI can also be read as 5.25^kT.

The completed force polygon of Figure A-21 is known as Maxwell's diagram.

PROBLEMS

A-10—A-21 Solve Problems 5-1 through 5-11 using the methods of Maxwell's diagram.

A-22 Using Maxwell's diagram, solve for the forces in all members of Figure P5-14.

A-23 Using Maxwell's diagram, solve for the forces in all members of Figure P5-15.

A-24 Using Maxwell's diagram, determine the loads in all members of the truss shown in Figure P5-21.

A-25 Using Maxwell's diagram, determine the loads in all members of the truss shown in Figure P5-22.

W SHAPES *
Properties for designing

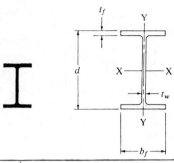

Designation	Area A	Depth d	Flange		Web Thick-ness t_w	Elastic Properties					
			Width b_f	Thick-ness t_f		Axis X-X			Axis Y-Y		
						I	S	r	I	S	r
	In.²	In.	In.	In.	In.	In.⁴	In.³	In.	In.⁴	In.³	In.
W 12×190	55.9	14.38	12.670	1.736	1.060	1890	263	5.82	590	93.1	3.25
×161	47.4	13.88	12.515	1.486	0.905	1540	222	5.70	486	77.7	3.20
×133	39.1	13.38	12.365	1.236	0.755	1220	183	5.59	390	63.1	3.16
×120	35.3	13.12	12.320	1.106	0.710	1070	163	5.51	345	56.0	3.13
×106	31.2	12.88	12.230	0.986	0.620	931	145	5.46	301	49.2	3.11
× 99	29.1	12.75	12.192	0.921	0.582	859	135	5.43	278	45.7	3.09
W 10×112	32.9	11.38	10.415	1.248	0.755	719	126	4.67	235	45.2	2.67
×100	29.4	11.12	10.345	1.118	0.685	625	112	4.61	207	39.9	2.65
× 89	26.2	10.88	10.275	0.998	0.615	542	99.7	4.55	181	35.2	2.63
× 77	22.7	10.62	10.195	0.868	0.535	457	86.1	4.49	153	30.1	2.60
× 72	21.2	10.50	10.170	0.808	0.510	421	80.1	4.46	142	27.9	2.59
× 66	19.4	10.38	10.117	0.748	0.457	382	73.7	4.44	129	25.5	2.58
× 60	17.7	10.25	10.075	0.683	0.415	344	67.1	4.41	116	23.1	2.57
× 54	15.9	10.12	10.028	0.618	0.368	306	60.4	4.39	104	20.7	2.56
× 49	14.4	10.00	10.000	0.558	0.340	273	54.6	4.35	93.0	18.6	2.54
W 8×67	19.7	9.00	8.287	0.933	0.575	272	60.4	3.71	88.6	21.4	2.12
×58	17.1	8.75	8.222	0.808	0.510	227	52.0	3.65	74.9	18.2	2.10
×48	14.1	8.50	8.117	0.683	0.405	184	43.2	3.61	60.9	15.0	2.08
×40	11.8	8.25	8.077	0.558	0.365	146	35.5	3.53	49.0	12.1	2.04
×35	10.3	8.12	8.027	0.493	0.315	126	31.1	3.50	42.5	10.6	2.03
×31	9.12	8.00	8.000	0.433	0.288	110	27.4	3.47	37.0	9.24	2.01
W 8×28	8.23	8.06	6.540	0.463	0.285	97.8	24.3	3.45	21.6	6.61	1.62
×24	7.06	7.93	6.500	0.398	0.245	82.5	20.8	3.42	18.2	5.61	1.61
W 8×20	5.89	8.14	5.268	0.378	0.248	69.4	17.0	3.43	9.22	3.50	1.25
×17	5.01	8.00	5.250	0.308	0.230	56.6	14.1	3.36	7.44	2.83	1.22
W 8×15	4.43	8.12	4.015	0.314	0.245	48.1	11.8	3.29	3.40	1.69	0.876
×13	3.83	8.00	4.000	0.254	0.230	39.6	9.90	3.21	2.72	1.36	0.842
×10	2.96	7.90	3.940	0.204	0.170	30.8	7.80	3.23	2.08	1.06	0.839
W 6×25	7.35	6.37	6.080	0.456	0.320	53.3	16.7	2.69	17.1	5.62	1.53
×20	5.88	6.20	6.018	0.367	0.258	41.5	13.4	2.66	13.3	4.43	1.51
×15.5	4.56	6.00	5.995	0.269	0.235	30.1	10.0	2.57	9.67	3.23	1.46

*Reprinted with the permission of the American Institute of Steel Construction, New York, N.Y., *Manual of Steel Construction*, Seventh Edition, 1970

S SHAPES *
Properties for designing

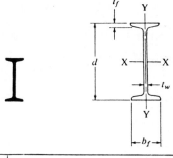

Designation	Area A	Depth d	Flange Width b_f	Flange Thickness t_f	Web Thickness t_w	Axis X-X I	Axis X-X S	Axis X-X r	Axis Y-Y I	Axis Y-Y S	Axis Y-Y r
	In.²	In.	In.	In.	In.	In.⁴	In.³	In.	In.⁴	In.³	In.
S 24×120	35.3	24.00	8.048	1.102	0.798	3030	252	9.26	84.2	20.9	1.54
×105.9	31.1	24.00	7.875	1.102	0.625	2830	236	9.53	78.2	19.8	1.58
S 24×100	29.4	24.00	7.247	0.871	0.747	2390	199	9.01	47.8	13.2	1.27
× 90	26.5	24.00	7.124	0.871	0.624	2250	187	9.22	44.9	12.6	1.30
× 79.9	23.5	24.00	7.001	0.871	0.501	2110	175	9.47	42.3	12.1	1.34
S 20× 95	27.9	20.00	7.200	0.916	0.800	1610	161	7.60	49.7	13.8	1.33
× 85	25.0	20.00	7.053	0.916	0.653	1520	152	7.79	46.2	13.1	1.36
S 20× 75	22.1	20.00	6.391	0.789	0.641	1280	128	7.60	29.6	9.28	1.16
× 65.4	19.2	20.00	6.250	0.789	0.500	1180	118	7.84	27.4	8.77	1.19
S 18× 70	20.6	18.00	6.251	0.691	0.711	926	103	6.71	24.1	7.72	1.08
× 54.7	16.1	18.00	6.001	0.691	0.461	804	89.4	7.07	20.8	6.94	1.14
S 15× 50	14.7	15.00	5.640	0.622	0.550	486	64.8	5.75	15.7	5.57	1.03
× 42.9	12.6	15.00	5.501	0.622	0.411	447	59.6	5.95	14.4	5.23	1.07
S 12× 50	14.7	12.00	5.477	0.659	0.687	305	50.8	4.55	15.7	5.74	1.03
× 40.8	12.0	12.00	5.252	0.659	0.472	272	45.4	4.77	13.6	5.16	1.06
S 12× 35	10.3	12.00	5.078	0.544	0.428	229	38.2	4.72	9.87	3.89	0.980
× 31.8	9.35	12.00	5.000	0.544	0.350	218	36.4	4.83	9.36	3.74	1.00
S 10× 35	10.3	10.00	4.944	0.491	0.594	147	29.4	3.78	8.36	3.38	0.901
× 25.4	7.46	10.00	4.661	0.491	0.311	124	24.7	4.07	6.79	2.91	0.954
S 8× 23	6.77	8.00	4.171	0.425	0.441	64.9	16.2	3.10	4.31	2.07	0.798
× 18.4	5.41	8.00	4.001	0.425	0.271	57.6	14.4	3.26	3.73	1.86	0.831
S 7× 20	5.88	7.00	3.860	0.392	0.450	42.4	12.1	2.69	3.17	1.64	0.734
× 15.3	4.50	7.00	3.662	0.392	0.252	36.7	10.5	2.86	2.64	1.44	0.766
S 6× 17.25	5.07	6.00	3.565	0.359	0.465	26.3	8.77	2.28	2.31	1.30	0.675
× 12.5	3.67	6.00	3.332	0.359	0.232	22.1	7.37	2.45	1.82	1.09	0.705
S 5× 14.75	4.34	5.00	3.284	0.326	0.494	15.2	6.09	1.87	1.67	1.01	0.620
× 10	2.94	5.00	3.004	0.326	0.214	12.3	4.92	2.05	1.22	0.809	0.643
S 4× 9.5	2.79	4.00	2.796	0.293	0.326	6.79	3.39	1.56	0.903	0.646	0.569
× 7.7	2.26	4.00	2.663	0.293	0.193	6.08	3.04	1.64	0.764	0.574	0.581
S 3× 7.5	2.21	3.00	2.509	0.260	0.349	2.93	1.95	1.15	0.586	0.468	0.516
× 5.7	1.67	3.00	2.330	0.260	0.170	2.52	1.68	1.23	0.455	0.390	0.522

*Reprinted with the permission of the American Institute of Steel Construction, New York, N.Y., *Manual of Steel Construction*, Seventh Edition, 1970

CHANNELS
AMERICAN STANDARD*
Properties for designing

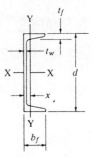

Designation	Area A	Depth d	Flange Width b_f	Flange Average thickness t_f	Web thickness t_w	$\dfrac{d}{A_f}$	Axis X-X I	Axis X-X S	Axis X-X r
	In.²	In.	In.	In.	In.		In.⁴	In.³	In.
C 15×50	14.7	15.00	3.716	0.650	0.716	6.21	404	53.8	5.24
×40	11.8	15.00	3.520	0.650	0.520	6.56	349	46.5	5.44
×33.9	9.96	15.00	3.400	0.650	0.400	6.79	315	42.0	5.62
C 12×30	8.82	12.00	3.170	0.501	0.510	7.55	162	27.0	4.29
×25	7.35	12.00	3.047	0.501	0.387	7.85	144	24.1	4.43
×20.7	6.09	12.00	2.942	0.501	0.282	8.13	129	21.5	4.61
C 10×30	8.82	10.00	3.033	0.436	0.673	7.55	103	20.7	3.42
×25	7.35	10.00	2.886	0.436	0.526	7.94	91.2	18.2	3.52
×20	5.88	10.00	2.739	0.436	0.379	8.36	78.9	15.8	3.66
×15.3	4.49	10.00	2.600	0.436	0.240	8.81	67.4	13.5	3.87
C 9×20	5.88	9.00	2.648	0.413	0.448	8.22	60.9	13.5	3.22
×15	4.41	9.00	2.485	0.413	0.285	8.76	51.0	11.3	3.40
×13.4	3.94	9.00	2.433	0.413	0.233	8.95	47.9	10.6	3.48
C 8×18.75	5.51	8.00	2.527	0.390	0.487	8.12	44.0	11.0	2.82
×13.75	4.04	8.00	2.343	0.390	0.303	8.75	36.1	9.03	2.99
×11.5	3.38	8.00	2.260	0.390	0.220	9.08	32.6	8.14	3.11
C 7×14.75	4.33	7.00	2.299	0.366	0.419	8.31	27.2	7.78	2.51
×12.25	3.60	7.00	2.194	0.366	0.314	8.71	24.2	6.93	2.60
× 9.8	2.87	7.00	2.090	0.366	0.210	9.14	21.3	6.08	2.72
C 6×13	3.83	6.00	2.157	0.343	0.437	8.10	17.4	5.80	2.13
×10.5	3.09	6.00	2.034	0.343	0.314	8.59	15.2	5.06	2.22
× 8.2	2.40	6.00	1.920	0.343	0.200	9.10	13.1	4.38	2.34
C 5× 9	2.64	5.00	1.885	0.320	0.325	8.29	8.90	3.56	1.83
× 6.7	1.97	5.00	1.750	0.320	0.190	8.93	7.49	3.00	1.95
C 4× 7.25	2.13	4.00	1.721	0.296	0.321	7.84	4.59	2.29	1.47
× 5.4	1.59	4.00	1.584	0.296	0.184	8.52	3.85	1.93	1.56
C 3× 6	1.76	3.00	1.596	0.273	0.356	6.87	2.07	1.38	1.08
× 5	1.47	3.00	1.498	0.273	0.258	7.32	1.85	1.24	1.12
× 4.1	1.21	3.00	1.410	0.273	0.170	7.78	1.66	1.10	1.17

*Reprinted with the permission of the American Institute of Steel Construction, New York, N.Y., *Manual of Steel Construction*, Seventh Edition, 1970

ANGLES
Equal legs[*]

Properties for designing

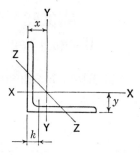

Size and Thickness	k	Weight per Foot	Area	AXIS X-X AND AXIS Y-Y				AXIS Z-Z
				I	S	r	x or y	r
In.	In.	Lb.	In.²	In.⁴	In.³	In.	In.	In.
L 8 × 8 × 1⅛	1¾	56.9	16.7	98.0	17.5	2.42	2.41	1.56
1	1⅝	51.0	15.0	89.0	15.8	2.44	2.37	1.56
⅞	1½	45.0	13.2	79.6	14.0	2.45	2.32	1.57
¾	1⅜	38.9	11.4	69.7	12.2	2.47	2.28	1.58
⅝	1¼	32.7	9.61	59.4	10.3	2.49	2.23	1.58
⁹⁄₁₆	1³⁄₁₆	29.6	8.63	54.1	9.34	2.50	2.21	1.59
½	1⅛	26.4	7.75	48.6	8.36	2.50	2.19	1.59
L 6 × 6 × 1	1½	37.4	11.0	35.5	8.57	1.80	1.86	1.17
⅞	1⅜	33.1	9.73	31.9	7.63	1.81	1.82	1.17
¾	1¼	28.7	8.44	28.2	6.66	1.83	1.78	1.17
⅝	1⅛	24.2	7.11	24.2	5.66	1.84	1.73	1.18
⁹⁄₁₆	1¹⁄₁₆	21.9	6.43	22.1	5.14	1.85	1.71	1.18
½	1	19.6	5.75	19.9	4.61	1.86	1.68	1.18
⁷⁄₁₆	¹⁵⁄₁₆	17.2	5.06	17.7	4.08	1.87	1.66	1.19
⅜	⅞	14.9	4.36	15.4	3.53	1.88	1.64	1.19
⁵⁄₁₆	¹³⁄₁₆	12.4	3.65	13.0	2.97	1.89	1.62	1.20
L 5 × 5 × ⅞	1⅜	27.2	7.98	17.8	5.17	1.49	1.57	.973
¾	1¼	23.6	6.94	15.7	4.53	1.51	1.52	.975
⅝	1⅛	20.0	5.86	13.6	3.86	1.52	1.48	.978
½	1	16.2	4.75	11.3	3.16	1.54	1.43	.983
⁷⁄₁₆	¹⁵⁄₁₆	14.3	4.18	10.0	2.79	1.55	1.41	.986
⅜	⅞	12.3	3.61	8.74	2.42	1.56	1.39	.990
⁵⁄₁₆	¹³⁄₁₆	10.3	3.03	7.42	2.04	1.57	1.37	.994
L 4 × 4 × ¾	1⅛	18.5	5.44	7.67	2.81	1.19	1.27	.778
⅝	1	15.7	4.61	6.66	2.40	1.20	1.23	.779
½	⅞	12.8	3.75	5.56	1.97	1.22	1.18	.782
⁷⁄₁₆	¹³⁄₁₆	11.3	3.31	4.97	1.75	1.23	1.16	.785
⅜	¾	9.8	2.86	4.36	1.52	1.23	1.14	.788
⁵⁄₁₆	¹¹⁄₁₆	8.2	2.40	3.71	1.29	1.24	1.12	.791
¼	⅝	6.6	1.94	3.04	1.05	1.25	1.09	.795

*Reprinted with the permission of the American Institute of Steel Construction, New York, N.Y., *Manual of Steel Construction*, Seventh Edition, 1970

ANGLES
Unequal legs[*]

Properties for designing

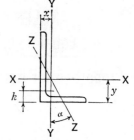

Size and Thickness	k	Weight per Foot	Area	AXIS X-X				AXIS Y-Y				AXIS Z-Z	
				I	S	r	y	I	S	r	x	r	Tan α
In.	In.	Lb.	In.²	In.⁴	In.³	In.	In.	In.⁴	In.³	In.	In.	In.	
L 9 × 4 × 1	1½	40.8	12.0	97.0	17.6	2.84	3.50	12.0	4.00	1.00	1.00	.834	.203
⅞	1⅜	36.1	10.6	86.8	15.7	2.86	3.45	10.8	3.56	1.01	.953	.836	.208
¾	1¼	31.3	9.19	76.1	13.6	2.88	3.41	9.63	3.11	1.02	.906	.841	.212
⅝	1⅛	26.3	7.73	64.9	11.5	2.90	3.36	8.32	2.65	1.04	.858	.847	.216
⁹⁄₁₆	1¹⁄₁₆	23.8	7.00	59.1	10.4	2.91	3.33	7.63	2.41	1.04	.834	.850	.218
½	1	21.3	6.25	53.2	9.34	2.92	3.31	6.92	2.17	1.05	.810	.854	.220
L 8 × 6 × 1	1½	44.2	13.0	80.8	15.1	2.49	2.65	38.8	8.92	1.73	1.65	1.28	.543
⅞	1⅜	39.1	11.5	72.3	13.4	2.51	2.61	34.9	7.94	1.74	1.61	1.28	.547
¾	1¼	33.8	9.94	63.4	11.7	2.53	2.56	30.7	6.92	1.76	1.56	1.29	.551
⅝	1⅛	28.5	8.36	54.1	9.87	2.54	2.52	26.3	5.88	1.77	1.52	1.29	.554
⁹⁄₁₆	1¹⁄₁₆	25.7	7.56	49.3	8.95	2.55	2.50	24.0	5.34	1.78	1.50	1.30	.556
½	1	23.0	6.75	44.3	8.02	2.56	2.47	21.7	4.79	1.79	1.47	1.30	.558
⁷⁄₁₆	¹⁵⁄₁₆	20.2	5.93	39.2	7.07	2.57	2.45	19.3	4.23	1.80	1.45	1.31	.560
L 6 × 4 × ⅞	1⅜	27.2	7.98	27.7	7.15	1.86	2.12	9.75	3.39	1.11	1.12	.857	.421
¾	1¼	23.6	6.94	24.5	6.25	1.88	2.08	8.68	2.97	1.12	1.08	.860	.428
⅝	1⅛	20.0	5.86	21.1	5.31	1.90	2.03	7.52	2.54	1.13	1.03	.864	.435
⁹⁄₁₆	1¹⁄₁₆	18.1	5.31	19.3	4.83	1.90	2.01	6.91	2.31	1.14	1.01	.866	.438
½	1	16.2	4.75	17.4	4.33	1.91	1.99	6.27	2.08	1.15	.987	.870	.440
⁷⁄₁₆	¹⁵⁄₁₆	14.3	4.18	15.5	3.83	1.92	1.96	5.60	1.85	1.16	.964	.873	.443
⅜	⅞	12.3	3.61	13.5	3.32	1.93	1.94	4.90	1.60	1.17	.941	.877	.446
⁵⁄₁₆	¹³⁄₁₆	10.3	3.03	11.4	2.79	1.94	1.92	4.18	1.35	1.17	.918	.882	.448
¼	¾	8.3	2.44	9.27	2.26	1.95	1.89	3.41	1.10	1.18	.894	.887	.451

*Reprinted with the permission of the American Institute of Steel Construction, New York, N.Y., *Manual of Steel Construction*, Seventh Edition, 1970

ANGLES
Unequal legs
Properties for designing

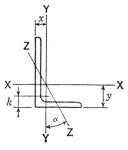

Size and Thickness	k	Weight per Foot	Area	AXIS X-X				AXIS Y-Y				AXIS Z-Z	
				I	S	r	y	I	S	r	x	r	Tan α
In.	In.	Lb.	In.²	In.⁴	In.³	In.	In.	In.⁴	In.³	In.	In.	In.	
L 4 × 3½ × ⅝	1¹⁄₁₆	14.7	4.30	6.37	2.35	1.22	1.29	4.52	1.84	1.03	1.04	.719	.745
½	¹⁵⁄₁₆	11.9	3.50	5.32	1.94	1.23	1.25	3.79	1.52	1.04	1.00	.722	.750
⁷⁄₁₆	⅞	10.6	3.09	4.76	1.72	1.24	1.23	3.40	1.35	1.05	.978	.724	.753
⅜	¹³⁄₁₆	9.1	2.67	4.18	1.49	1.25	1.21	2.95	1.17	1.06	.955	.727	.755
⁵⁄₁₆	¾	7.7	2.25	3.56	1.26	1.26	1.18	2.55	.994	1.07	.932	.730	.757
¼	¹¹⁄₁₆	6.2	1.81	2.91	1.03	1.27	1.16	2.09	.808	1.07	.909	.734	.759
L 4 × 3 × ⅝	1¹⁄₁₆	13.6	3.98	6.03	2.30	1.23	1.37	2.87	1.35	.849	.871	.637	.534
½	¹⁵⁄₁₆	11.1	3.25	5.05	1.89	1.25	1.33	2.42	1.12	.864	.827	.639	.543
⁷⁄₁₆	⅞	9.8	2.87	4.52	1.68	1.25	1.30	2.18	.992	.871	.804	.641	.547
⅜	¹³⁄₁₆	8.5	2.48	3.96	1.46	1.26	1.28	1.92	.866	.879	.782	.644	.551
⁵⁄₁₆	¾	7.2	2.09	3.38	1.23	1.27	1.26	1.65	.734	.887	.759	.647	.554
¼	¹¹⁄₁₆	5.8	1.69	2.77	1.00	1.28	1.24	1.36	.599	.896	.736	.651	.558
L 3½ × 3 × ½	¹⁵⁄₁₆	10.2	3.00	3.45	1.45	1.07	1.13	2.33	1.10	.881	.875	.621	.714
⁷⁄₁₆	⅞	9.1	2.65	3.10	1.29	1.08	1.10	2.09	.975	.889	.853	.622	.718
⅜	¹³⁄₁₆	7.9	2.30	2.72	1.13	1.09	1.08	1.85	.851	.897	.830	.625	.721
⁵⁄₁₆	¾	6.6	1.93	2.33	.954	1.10	1.06	1.58	.722	.905	.808	.627	.724
¼	¹¹⁄₁₆	5.4	1.56	1.91	.776	1.11	1.04	1.30	.589	.914	.785	.631	.727

ANSWERS TO
ODD-NUMBERED PROBLEMS

Chapter 1

1-1 23 in., 19.3 in.
1-3 R = 5 in., θ = 36.9°
R = 13 in., θ = 67.4°
R = 17 in., θ = 28°
1-5 θ = 38.7°
1-7 CB = 46.2″
1-9 15.9°, 45.3°, 118.8°
1-11 h = 3.57 in.
1-13 AC = 16.5 m, BC = 13.4 m
1-15 6.49 cm
1-17 y = 6.34 m, x = 13.6 m
1-19 0.79 m
1-21 d = 4.85 m x = 1.95 m

Chapter 2

2-1 y = 4.62 in.
2-3 y = 59.6 mm, x = 110 mm, d = 43.3 mm
2-5 R = 65 N $^{12}\!\!\diagdown_5$, R = 8.54 kN $\diagup^{8}_{3}$, R = 102 N $\diagup^{15}_{8}$
2-7 R = 17 lb $\diagup^{15}_{8}$
2-9 R = 1.52 kN $\diagdown$ 29.5°

445

2-11 $F_x = 11$ N $\rightarrow$ $\qquad$ $F_y = 38.4$ N $\downarrow$
$\qquad$ $F_x = 61.3$ kN $\leftarrow$ $\qquad$ $F_y = 51.4$ kN $\downarrow$
$\qquad$ $F_x = 564$ N $\leftarrow$ $\qquad$ $F_y = 205$ N $\uparrow$
$\qquad$ $F_x = 6$ kN $\rightarrow$ $\qquad$ $F_y = 10.4$ kN $\uparrow$
2-13 $F_x = 52$ lb $\rightarrow$ $\qquad$ $F_y = 30$ lb $\uparrow$
$\qquad$ $F_x = 27.4$ kips $\leftarrow$ $\qquad$ $F_y = 75.2$ kips $\downarrow$
$\qquad$ $V_x = 300$ ft/sec $\leftarrow$ $\qquad$ $V_y = 400$ ft/sec $\uparrow$
$\qquad$ $V_x = 30$ mph $\leftarrow$ $\qquad$ $V_y = 16$ mph $\downarrow$
2-15 $F_x = 24$ lb $\nearrow$, $F_y = 18$ lb $\nwarrow$
2-17 $A_x = 451$ N $\rightarrow$, $A_y = 451$ N $\downarrow$
2-19 $F_x = 10$ lb $\nearrow$
2-21 $R = 62.6$ N $\quad \underset{3}{\overset{10}{\angle}}$
2-23 $R = 9.7$ kips $\quad \diagup 85.6°$
2-25 $R = 33.1$ N $\underline{25°\nwarrow}$
2-27 $R = 459$ N $\underline{23°\nwarrow}$
2-29 $R = 63.5$ lb

Chapter 3

3-1 $M_A = 1$ lb-ft $\curvearrowright$
3-3 $M_C = 402$ lb-ft $\curvearrowright$
3-5 $M_A = 1280$ lb–ft $\curvearrowleft$
3-7 $M_A = 2160$ N·m $\curvearrowright$
3-9 $M_P = 420$ lb-in. $\curvearrowright$
3-11 $M_A = 3000$ N·m $\curvearrowright$
$\qquad$ $M_B = 1300$ N·m $\curvearrowright$
3-13 $\underline{40 \text{ lb}\downarrow \quad 4' \quad \uparrow 40 \text{ lb}}$
3-15 $\begin{array}{l} \overset{\longleftarrow \quad 53.3 \text{ N}}{\underset{\longrightarrow 53.3 \text{ N}}{|\; 150 \text{ mm}}} \end{array}$
3-17 $\begin{array}{l} 160 \text{ N} \downarrow \qquad \uparrow 160 \text{ N} \\ \overline{\quad 50 \text{ mm} \qquad\qquad} | \end{array}$
3-19 $M_A = M_B = 169$ kip-ft $\curvearrowright$
3-21 $M_A = M_B = 45$ N·m $\curvearrowright$

Chapter 4

4-19 $T_A = 20$ lb, $T_B = 20$ lb
4-21 75 lb
4-23 300 lb
4-25 AC $= 1000$ N T, $\quad$ BC $= 800$ N C
4-27 AB $= 75$ lb T $\qquad$ BC $= 125$ lb T
$\qquad$ CD $= 241$ lb T $\qquad$ CE $= 160$ lb T
4-29 AB $= 750$ N T $\qquad$ BC $= 770$ N C
4-31 AC $= 7.26$ kN C $\quad$ AB $= 4$ kN T
4-33 $R_A = 31.7$ lb $\rightarrow$ $\qquad$ $R_B = 50$ lb $\uparrow$
4-35 AD $= 2.45$ kN T $\qquad$ BD $= 2.19$ kN C
$\qquad$ DE $= 1.79$ kN T $\qquad$ CE $= 0.53$ kN T

4-37 0.97 kN
4-39 A = 19 lb↓   B = 169 lb↑
4-41 A = 0.89 kN↑   B = 1.01 kN↑
4-43 A = 9 kN↑   B = 9.5 kN↑
4-45 A = 86.5 kN↑   B = 116 kN↑
4-47 A = 3.28 kN↑   B = 5.22 kN↑
4-49 A = 7.9 kips↑   B = 9.1 kips↑
4-51 B = 523 lb C   C = 571 lb T
4-53 A = 4380 lb↑   B = 3530 lb↑   C = 3090 lb↑
4-55 A_x = 191 lb← A_y = 374 lb↓   B = 281 lb 4⟍₃
4-57 B_x = 3360 N→ B_y = 600 N↓   C = 3040 N←
4-59 A = 28.6 lb→ B_x = 28.6 lb← B_y = 400 lb↑
4-61 A_x = 200 lb← A_y = 215 lb↓   B = 215 lb↑
4-63 A = 70 N↓   B = 10 N↓   C = 70 N↑   D = 10 N↑

Chapter 5

5-1 AB = 10.6 kN T   BC = 6.37 kN T
      BD = 8.5 kN C   AD = 6.35 kN C
      DC = 7.2 kN C
5-3 AB = 0   AC = 2.25 kips T CD = 3 kips T
      CE = 4.5 kips T   CG = 3.75 kips C DG = 6.25 kips C
      EH = 9.83 kips T   EJ = 6.67 kips C EG = 3 kips T
      GJ = 8.5 kips C   BD = 4 kips C   AD = 3.75 kips C
5-5 AB = 14.4 kips C BC = 11 kips C   CD = 9.82 kips C
      DE = 9.82 kips C EG = 4.91 kips T HG = 14.7 kips T
      AH = 7.21 kips T BH = 7.51 kips T CH = 7.51 kips C
      CG = 9.82 kips C DG = 9.82 kips T
5-7 AB = 77.8 kN T   BC = 36 kN T   AC = 30 kN C
      DB = 79 kN C
5-9 AB = 14.1 kips T BC = 10.5 kips T CD = 16.7 kips C
      BD = 13.3 kips T EB = 0   ED = 10 kips C
5-11 AB = 5.25 kN C   BC = 6.67 kN C CD = 6.67 kN C
      DE = 6 kN C   BE = 3.55 kN T AD = 5.69 kN T
5-13 3000 lb T
5-15 BC = 6400 N T   BG 4000 N C   EG = 3200 N C
5-17 BC = 6.67 kips C BE = 1.94 kips C
      BD = 0   DE = 8.33 kips T
5-19 CD = 83.3 kN T   EG = 41.7 kN C EH = 33.3 kN C
5-21 CD = 0.75 kN C   ED = 3.36 kN C
5-23 DE = 7.78 kips C GE = 0.71 kips C GH = 2 kips T
5-25 CD = 2.4 kips C   CG = 3 kips C
5-27 DE = 3.39 kN C   JE = 0.81 kN T KH = 2.25 kN T
5-29 B_y = 0 B_x = 7.5 kips D_y = 5 kips D_x = 7.5 kips
5-31 B_x = 200 lb B_y = 167 lb C_x = 300 lb C_y = 67 lb
5-33 B_x = 0 B_y = 16.5 kN C_x = 3 kN C_y = 5.5 kN

5-35 $B_x = 6.67$ kN← $B_y = 5$ kN↑
     $C_x = 4.67$ kN→ $C_y = 5$ kN↓
5-37 $A_x = 62.5$ lb $A_y = 375$ lb
     $D_x = 250$ lb $D_y = 375$ lb
5-39 $P = 5.21$ kN
     $A_x = 2$ kN $A_y = 6.67$ kN
     $B_x = 5$ kN $B_y = 1.46$ kN
5-41 $DH = 27.1$ kips T
5-43 $DE = 359$ N T
5-45 $EH = 750$ lb
5-47 $D = 7.87$ kN
     $A_x = 2.19$ kN $A_y = 0$
     $E_x = 2.19$ kN $E_y = 4.37$ kN
5-49 $BG = 735$ lb C
5-51 $A = 396$ lb ⁴◺₃ $B_x = 15$ lb ◿³₄ $B_y = 360$ lb ⁴◺₃
5-53 $A_x = 3750$ lb $A_y = 3875$ lb
     $B_x = 5750$ lb $B_y = 3375$ lb
5-55 $C_x = 675$ N $C_y = 2620$ N
     $D_x = 675$ N $D_y = 525$ N
5-57 $A_x = 8$ kips $A_y = 12$ kips

Chapter 6

6-1 $R = 50$ N↓ $\bar{x} = 6$ m $\bar{z} = 1.8$ m
6-3 $R = 1$ kip↓ $\bar{x} = -7$ ft $\bar{z} = -24$ ft
6-5 $R = 1.3$ kN↓ $\bar{x} = -4$ m $\bar{z} = 0.38$ m
6-7 $A = 0.6$ kips↑ $B = 3.38$ kips↑ $C = 2.57$ kips↑
6-9 $A = 3.59$ kN↑ $B = 2.26$ kN↑ $C = 0.15$ kN↑
6-11 $A = 1760$ lb↑ $B = 1180$ lb↑ $C = 2060$ lb↑
6-13 $R = 30.8$ kN $(3, 2, 5)$
6-15 $R = 16.6$ lb $(9, 5, 13)$
6-17 $R_x = 13.2$ N← $R_y = 46.4$ N↓ $R_z = 13.2$ N↙
6-19 $R_x = 244$ lb→ $R_y = 142$ lb↑ $R_z = 102$ lb↙
6-21 $R_x = 133$ lb← $R_y = 665$ lb↓ $R_z = 177$ lb↙
6-23 $R = 16.1$ kN
6-25 $R = 186$ lb
6-27 $C_x = 5.9$ lb $C_y = 6.1$ lb
     $D_x = 7.1$ lb $D_y = 37.6$ lb
6-29 $AC = 4.8$ kips T $BC = 4.8$ kips T $CD = 18.5$ kips C
6-31 $AB = 3.1$ kN T $BC = 4.27$ kN C $BD = 2.6$ kN T
6-33 $AB = 3.97$ kips C $CB = 3.46$ kips C $BD = 5.92$ kips C
6-35 $BC = 370$ lb T $BD = 295$ lb T
6-37 $AB = 3.35$ kN C $CB = 3.33$ kN T $DB = 1.98$ kN T
6-39 $AB = 2.06$ kN T $BC = 2.08$ kN T $HK = 0.447$ kN T
     $D_x = 0.16$ kN $D_y = 0.312$ kN $D_z = 2.57$ kN

Chapter 7

7-1 0.25
7-3 8 lb↗ 35 lb↙
7-5 33.3 N→
7-7 No sliding
7-9 7.6 lb→
7-11 P=50 lb, sliding
7-13 0.375
7-15 P=3180 lb
7-17 P=289 N
7-19 P=4.4 lb
7-21 P=1.44 lb
7-23 N_B=1680 N N_A=1410 N μ_A=0.24
         F_B=504 N F_A=342 N
7-25 9.56 Kg
7-27 No slipping at A
         min F_A=13 lb
         μ=0.139
7-29 58.1 lb
7-31 T_L=1030 lb T_S=243 lb
7-33 μ=0.368
7-35 41.2 lb
7-37 θ=35°
7-39 3.87 lb-ft
7-41 23.5 N

Chapter 8

8-1 $\bar{y}$=4.81 in.
8-3 $\bar{x}$=4.02 in. $\bar{y}$=5.96 in.
8-5 $\bar{x}$=0 $\bar{y}$=3.57 in.
8-7 $\bar{x}$=224 mm $\bar{y}$=50.3 mm
8-9 $\bar{x}$=48.6 mm $\bar{y}$=180 mm
8-11 $\bar{x}$=2.96 in. $\bar{y}$=2.92 in.
8-13 $\bar{y}$=84.6 mm
8-15 $\bar{y}$=57.5 mm above base
8-17 $\bar{x}$=6.67 in. $\bar{y}$=0.49 in.
8-19 $\bar{x}$=1 m $\bar{y}$=0.36 m
8-21 $\bar{y}$=75.7 mm

Chapter 9

9-1 2.13×10^8 mm^4
9-3 24.3×10^6 mm^4
9-5 369 in.^{4}

9-7 2.88×10^{-4} m^4
9-9 9.38×10^6 mm^4
9-11 834 in^4
9-13 95.1×10^6 mm^4
9-15 7.6×10^{-6} m^4
9-17 114×10^{-6} m^4
9-19 242 in^4
9-21 34.5×10^6 mm^4
9-23 53.7×10^6 mm^4
9-25 87.7×10^6 mm^4
9-27 11×10^6 mm^4
9-29 3.75×10^6 mm^4
9-31 99.9 in^4
9-33 535×10^6 mm^4
9-35 40.3 mm
9-37 0.185 m
9-39 $I = 0.18$ kg·m^2 $k = 0.0424$ m
9-41 0.397 kg·m^2
9-43 13.8 ft-lb-sec^2
9-45 0.083 kg·m^2
9-47 0.0376 kg·m^2

Chapter 10

10-1 displ. $= 7.16$ m   dist. $= 9.5$ m
10-3 7.54 ft
10-5 2.4 m/s
10-7 speed $= 140$ m/min   velocity $= 0$
10-9 30 ft/sec
10-11 1.22 m/s^2 $\angle 67.5°$
10-13 0.75 m/s^2
10-15 $v_0 = 11$ ft/s $v = 41$ ft/s
10-17 5 m/s^2
10-19 $t = 3.5$ s $v = 34.3$ m/s
10-21 5.64 ft/sec   13.9 ft/sec
10-23 193 ft/sec   299 sec
10-25 37.8 m/s $\nearrow 31.6°$
10-27 426 m   163 m/s $\overline{29.9°}$
10-29 $d = 1$ ft

Chapter 11

11-1 2π rad π rad
11-3 96.6 rad
11-5 9.66 rad

11-7 2100 rpm
11-9 18,000 rad/min²   2870 rev/min²
11-13 0.91 rad/s²   33 s
11-15 6.98 rad/s²   2.59 rev
11-17 22.8 min
11-19 0.26 rad/s²
11-21 24 s   3.48 rad/s²
11-23 4.51 s   4.18 rad/s²   17.3 rev
11-25 1.74 rad/s   3.49 rad/s²
11-27 840 m
11-29 16.7 m/s
11-31 3.77 m/s   9.42 rad/s
11-33 27.2 m/s
11-35 62.8 in./sec   200 rpm   300 rpm
11-37 808 rpm   62.4 mph
11-39 1.2 m/s²   4.8 m/s
11-41 11.1 rad/s²   1.33 m/s²
11-43 $\theta_B = 100$ rad $\downarrow$ $\theta_C = 267$ rad $\uparrow$
    $\alpha_B = 0.5$ rad/s² $\downarrow$ $\alpha_C = 1.33$ rad/s² $\uparrow$
    $a_t = 0.33$ ft/sec²
11-45 231 rad/s
11-47 14.4 m/s²
11-49 52.8 mph
11-51 28.3 m/s
11-53 8.69 ft/sec²

Chapter 12

12-1 11.7 ft
12-3 $s_{A/B} = 6$ m $\downarrow$
12-5 $v_{A/B} = 13.5$ m/s 65.8° $\nwarrow$
12-7 $v_{A/B} = 110$ mph $\uparrow$ $v_{B/A} = 110$ mph $\downarrow$
12-9 $s_{B/A} = 215$ m $\triangle$₂ $v_{B/A} = 53.8$ m/s $\triangle$₂ $a_{B/A} = 6.7$ m/s² $\triangle$₂
12-11 $v_C = 7.33$ m/s $\rightarrow$ $\omega_{BC} = 0$
    $v_C = 1.11$ m/s $\omega_{BC} = 14.8$ rad/s $\uparrow$
12-13 $v_D = 2.88$ m/s $\rightarrow$ $v_E = 4.8$ m/s $\leftarrow$
12-15 50.8 in./sec²
12-17 28.2 m/s² 1.5° $\swarrow$
12-19 $\omega = 7.5$ rad/s $\downarrow$ $v_B = 8.48$ m/s $\nearrow$45°
12-21 $v_B = 38.6$ ft/sec 15° $\nwarrow$
12-23 $v_D = 14.8$ ft/sec $\triangle$¹⁵ $v_E = 20$ ft/sec $\rightarrow$
    $v_c = 6.95$ ft/sec $\rightarrow$ $s_C = 3.48$ ft
12-25 17.8 m
12-27 4 rad/s²
12-29 347 rad/s $\downarrow$

12-31 $v_A = 33.3$ m/s $\rightarrow$

      $v_B = 25.4$ m/s $^7\!\searrow_3$ $v_D = 13.3$ m/s $\leftarrow$

12-33 3.28 ft/min $\rightarrow$

12-35 $v_A = 16.1$ ft/sec $\downarrow$ $v_C = 13.3$ ft/sec $\swarrow^{17}_{3}$

12-37 $v_D = 3.54$ m/s $\measuredangle^5_{12}$

12-39 246 mm/s $\nearrow 30°$

12-41 $v_E = 272$ mm/s $\measuredangle^8_{15}$ $v_C = 338$ mm/s $\searrow 45°$

12-43 $\omega = 0.6$ rad/s $\circlearrowright$ $v_B = 0.849$ m/s $\searrow 45°$

12-45 $v_B = 700$ mm/s $\uparrow$ $\omega_{AB} = 1.87$ rad/s $\circlearrowright$

Chapter 13

13-1 $a = 4.29$ ft/sec^2 $\rightarrow$

13-3   239 N   121 N

13-5   264 lb

13-7   5.38 ft/sec^2 $\rightarrow$

13-9 $a_A = 1.4$ m/s^2 $\uparrow$ $a_B = 2.8$ m/s^2 $\downarrow$

13-11 $a_A = 4.9$ m/s^2 $\uparrow$ $a_B = 2.45$ m/s^2 $\downarrow$

13-13 6.18 m/s $\nearrow$

13-15 16.1 ft/sec

13-17 2270 N

13-19 $t = 46.5$ N $\theta = 32.5°$ $AB = 1.86$ m

13-21 1.15 rad/s

13-23 33.6 kg

13-25 0.911 m

13-27 $d = 4.06$ m $v = 2.29$ m/s $\swarrow$

13-29 34.9 s

13-31 4.52 rad/s^2 $\circlearrowright$

13-33 1.38 lb-ft

13-35 3000 N   6.07 rad/s^2 $\circlearrowright$

13-37 2260 N   8.14 rad/s^2 $\circlearrowright$

13-39 4.71 s

13-41 $A_x = 25$ N $\rightarrow$ $A_y = 44.1$ N $\uparrow$

13-43 $\alpha = 28.64$ rad/s^2 $\circlearrowright$ $A_x = 150$ N $\leftarrow$ $A_y = 137$ N $\uparrow$

13-45 $P = 86.2$ N $_3\searrow_4$

13-47 1.89 m/s^2 $\rightarrow$

13-49 6.84 m/s^2 $\rightarrow$

13-51 $P = 138$ lb $\downarrow$ $a = 9.2$ ft/sec^2

13-53 141 N   72.6 m

13-55 $\mu = 0.5$

13-57 5.7 rad/s^2 $\circlearrowright$

Chapter 14

14-1 120 ft-lb
14-3 7.17 kJ
14-5 2120 ft-lb
14-7 180 ft-lb
14-9 2.36 kJ
14-11 750 ft-lb
14-13 392 J
14-15 20 lb/in.   160 in.-lb
14-17 2500 N/m
14-19 20.4 kN/m
14-21 840 J
14-23 240 in.-lb
14-25 2.17 kJ   714 J
14-27 1.28 MJ 31.9 m
14-29 1.96 kJ   8.85 m/s
14-31 3.2 kJ   7.2 kJ
14-33 1.9 m/s↓
14-35 16.1 ft/sec
14-37 5.53 m
14-39 83.3 kN
14-41 5.14 ft/sec
14-43 0.127 m
14-45 15.5 ft/sec
14-47 2.36 m/s
14-49 15.3 ft/sec
14-51 6.8 ft-lb
14-53 45.5 J
14-55 62.8 J
14-57 239 lb-ft
14-59 1.04 m
14-61 11.6 ft/sec↑
14-63 18 ft
14-65 4.39 ft/sec↓
14-67 37,700 ft-lb   18,800 ft-lb
14-69 32.1 ft/sec
14-71 8.09 m/s↓
14-73 7.15 rad/s↻
14-75 36,800 ft lb/sec
14-77 23.5 MJ
14-79 6000 hp
14-81 129 hp
14-83 33.5 kW
14-85 39.8 hp

Chapter 15

15-1 5400 N·S
15-3 322 ft/sec
15-5 0.14 sec
15-7 7.69 m/s   15.4 m/s
15-9 14.6 m/s
15-11 6.93 s
15-13 16.1 ft/sec
15-15 18.8 s
15-17 396 N
15-19 1.03×10^{-2} N·m
15-21 35.6 ft-lb-sec^2
15-23 1.04 s
15-25 67.8 lb
15-27 181 rad/sec
15-29 2.86 m/s
15-31 2.46 m/s ⟍24°↗
15-33 1.27 s
15-35 35.5 rpm
15-37 70.1 kg

INDEX